Transition

Sandeep Chavan

Published by Gyrus Vision, 2026.

TRANSITION

First edition. March 10, 2026.

ISBN: 979-8233103469

Written by Sandeep Chavan.

Table of Contents

To every curious mindthat looks at the world and asks,

"How does this work?"

Because every discovery in human history began with a simple question.

"The universe has always been speaking.Human civilization is the story of learning how to listen."

Opening Note

A Question That Stayed With Me

This book did not begin in a laboratory, a library, or while reading a research paper. It began in a classroom.

I was teaching crystallography that day. The topic was crystal defects—vacancies, interstitials, non-stoichiometric compounds, oxidation states, charge neutrality. The language of solid-state chemistry was doing what it always does: describing the beauty of order in matter through careful definitions and mathematical clarity.

Students were following the lecture, some writing notes, some watching the diagrams forming on the board.

Then a student raised his hand.

He asked a question that stopped the entire flow of the lecture.

"Sir, the universe seems to run so simply. Then why do we describe it in such complicated, mathematical language?"

For a few seconds the class became completely silent.

It was not a challenge. It was not confusion. It was genuine curiosity.

And I realized that the student had unknowingly touched a very deep question.

Why do we explain nature through such structured scientific language when the world around us appears so effortless? Why do equations, models, and technical terminology stand between a human mind and the simple phenomena it observes every day?

That question stayed with me long after the lecture ended.

As a teacher, I often explain ideas in two different ways. One language is intuitive—the language of curiosity, stories, and visualization. The other language is scientific—the language of measurements, models, equations, and precision.

Both describe the same universe.

But they belong to different stages of human understanding.

Human civilization did not begin with equations. It began with stories told around fire. Those stories slowly transformed into questions. Questions evolved into philosophy. Philosophy matured into science. Science eventually became engineering and technology.

In other words, everything we call modern knowledge is the result of a long intellectual transition.

This book is about that transition.

It is about how a curious human sitting outside a cave slowly became a philosopher, a scientist, and an engineer. It is about how simple observations grew into powerful explanations. It is about why mathematics became the language of nature—and why behind that language there still exists a simple human curiosity trying to understand the world.

The goal of this book is not to make science more complicated. It is to show how complexity grows naturally from simple questions.

If this book succeeds, you will begin to see something remarkable: science, philosophy, and engineering are not separate domains. They are different stages of the same human journey of understanding.

And that journey began with a simple question.

A question asked by a student in a classroom.

Sometimes the most powerful ideas do not begin in laboratories.

They begin when someone looks at the universe with honest curiosity and simply asks,

"Why?"

Preface

Human civilization did not begin with science.
It began with curiosity.

Long before laboratories, universities, and scientific instruments existed, human beings were already observing the world around them. They watched the movement of the stars, felt the warmth of fire, experienced storms and seasons, and wondered about life and death. These early experiences raised questions that slowly shaped the intellectual journey of humanity.

At first, people explained nature through stories and myths. These stories were attempts to understand the mysterious forces surrounding human life. Over time, however, curiosity grew stronger and questions became deeper. Thinkers began to search for patterns in nature and to reason about the causes behind natural events.

This search gradually gave birth to philosophy.

Later, observation and experimentation transformed philosophical reflection into science. By carefully measuring and testing ideas, humans discovered that the universe follows consistent laws governing motion, energy, electricity, light, and matter.

Engineering then carried this knowledge into practical reality.

The scientific principles discovered over centuries became the foundation for machines, infrastructure, communication systems, and technologies that define modern civilization. The journey that began with simple curiosity eventually produced the technological world we live in today.

This book tells the story of that journey.

It explores how human understanding evolved—from stories to philosophy, from science to engineering, and finally to the technologies that shape our daily lives. The purpose of the book is not to present complex scientific theories, but to show the broader pattern through which knowledge develops.

As a teacher, I have often seen students feel overwhelmed by the complexity of scientific language and mathematical expressions. Yet behind those formal descriptions lies a very simple and human process: observing the world, asking questions, testing ideas, and gradually learning how nature works.

Understanding this journey helps us see science not as a collection of difficult subjects, but as a natural extension of human curiosity.

The same curiosity that once led our ancestors to watch the stars continues to guide modern research into atoms, galaxies, and consciousness. Humanity is still learning how to interpret the patterns of the universe.

This book invites readers to reflect on that ongoing journey—and perhaps to see themselves as part of it.

Prologue

The Night Outside the Cave

Night had fallen slowly.

The sun had disappeared beyond the hills, leaving behind a fading orange glow that slowly dissolved into darkness. A small fire burned outside the entrance of a cave. Sparks occasionally rose into the air and disappeared into the vast sky above.

A human sat near the fire.

There were no books, no instruments, no language for science yet. Only the raw experience of existence. The warmth of the fire on the skin. The distant sound of animals moving through the forest. The rhythm of breath. The quiet presence of the night.

Above him stretched the sky.

Thousands of small lights were scattered across the darkness. Some appeared fixed. Some slowly moved across the night. Occasionally a bright streak would cut across the sky and vanish.

The human did not know what stars were.

He did not know about galaxies, nuclear fusion, gravity, or cosmic expansion. He only knew that the lights appeared every night and disappeared every morning.

And that was enough to create wonder.

A gust of wind moved through the trees.

The fire flickered and changed shape. Flames danced, shadows shifted across the rocks, and the cave walls briefly came alive with movement. Fire was mysterious. It gave warmth, protection, and light, yet it could also burn and destroy.

The human did not know the chemistry of combustion.

But he knew fire mattered.

Somewhere nearby thunder rolled across the sky. A storm was approaching. Lightning flashed briefly, illuminating the entire valley for a moment before darkness returned again.

Storms were frightening.

They came suddenly, shook the sky, brought rain, and then disappeared as mysteriously as they had arrived.

Life itself was equally mysterious.

Children were born. People grew older. Animals hunted and were hunted. Sometimes people fell sick and died without any clear reason.

The human sitting outside the cave could not explain any of it.

But something important was happening inside his mind.

Questions were forming.

Why does fire burn?

Why does the sun return every morning?

Why do storms appear and disappear?

Why do people live and die?

These questions were not written anywhere. They were not discussed in schools or laboratories. They existed silently inside human awareness.

The earliest humans did not have the tools to investigate nature scientifically. So the first explanations came in the form of stories. Thunder became the voice of powerful beings. The sun became a traveler crossing the sky. Illness became the influence of unseen forces.

Stories were not mistakes.

They were the first attempts of the human mind to bring order to a mysterious universe.

But curiosity did not stop there.

Over time humans began noticing patterns. The sun always rose from the same direction. The moon changed shape in a repeating cycle. Seeds placed in soil eventually grew into plants. Fire always behaved in certain ways.

Patterns slowly replaced pure imagination.

Stories slowly transformed into questions.

Questions slowly became observation.

And observation slowly gave birth to something new: the desire to understand the universe not only through imagination, but through reasoning.

That moment marked the beginning of civilization.

Every scientific theory, every machine, every technology that surrounds us today can be traced back to that simple scene — a human sitting outside a cave, watching the fire, looking at the stars, and wondering how the world works.

Civilization did not begin with equations.

It began with curiosity.

And curiosity, once awakened, never stops asking questions.

PART I — STORIES

"The first tool of civilization was not a weapon or a machine — it was a question."

Chapter 1 — The First Language of the Universe

How early humans explained nature through stories, myths, and imagination

Imagine a night thousands of years ago.

A small group of humans sits around a fire. The flames move and crackle, throwing light onto rough stone walls and tired faces. The warmth of the fire keeps the cold away, but beyond the circle of light lies a vast darkness filled with sounds that no one fully understands.

Above them stretches the sky.

The stars shine with quiet persistence. They appear every night, arranged in patterns that seem almost intentional. Sometimes a bright streak cuts across the darkness and disappears. Sometimes the moon grows full and bright, and other nights it becomes only a thin curve before vanishing again.

To the people sitting by the fire, these are not astronomical objects.

They are mysteries.

A sudden flash of lightning appears far away on the horizon. Moments later thunder rolls across the land, shaking the silence. The children move closer to the fire. The elders look toward the sky, wondering what power lives above them.

No one in that circle knows about electricity, atmospheric charge, or planetary motion. They do not know about gravity holding celestial bodies in place or nuclear reactions inside distant stars.

Yet their minds are not empty.

The human mind cannot tolerate complete mystery for long. When it encounters the unknown, it begins to weave meaning.

Someone says the thunder is the voice of a powerful being in the sky. Another says the stars are fires lit by ancestors who now live above the world. The sun, which disappears every evening and returns every morning, becomes a traveler crossing the heavens.

These explanations are not experiments or theories.

They are stories.

Stories were the first language humans used to speak about the universe. Long before mathematics, instruments, and laboratories, the human mind used imagination to organize what it saw. A story could give shape to fear, comfort the mind, and create a sense of order in a world that seemed unpredictable.

These early stories were not signs of ignorance. They were signs of something far more important: the awakening of curiosity.

The human mind had begun asking questions.

Why does the sun rise and set? Why do storms come and go? Why do people live and die? Why do the stars move slowly across the night sky?

At first, the answers came as myths and legends. Over time, however, something remarkable happened. The same curiosity that created stories would eventually lead humans to seek deeper explanations.

The first language of the universe was storytelling.

But hidden inside every story was the beginning of a question.

1.1 — When Observation Met Imagination

Long before science existed, humans were already excellent observers.

Survival demanded it.

Early humans watched the movement of animals, the direction of the wind, the patterns of rainfall, and the changing cycles of seasons. A person who could not notice these patterns would struggle to hunt, gather food, or protect the group from danger. Observation was not an academic activity—it was a matter of life and survival.

Day after day, humans witnessed the same natural events repeating around them. The sun rose from one side of the horizon and disappeared on the other. The moon changed shape over many nights. Rivers flowed downhill. Fire produced heat and light. Plants grew from seeds placed in soil.

These patterns were noticed carefully.

But noticing something is not the same as understanding it.

The early human mind had no scientific framework to explain these events. There were no instruments to measure time precisely, no mathematics to describe motion, and no experiments designed to test ideas about nature.

Yet the mind still needed explanations.

This is where imagination entered the picture.

Whenever humans observed something they could not explain, the mind filled the gap with meaning. Thunder became the voice of a powerful sky-being. The sun became a divine traveler moving across the heavens. The moon's changing shape became a story about growth and disappearance.

These explanations may appear primitive from a modern scientific perspective, but they served an important purpose. They helped transform a chaotic and unpredictable world into something emotionally understandable.

Observation provided the raw experience of nature.

Imagination provided the narrative that made those experiences meaningful.

Together, observation and imagination formed the earliest way humans interpreted the universe.

This combination is important to understand because imagination was not the enemy of knowledge. In fact, imagination was a necessary first step. It allowed humans to construct mental models of the world before tools existed to test them.

In many ways, imagination acted as the first bridge between curiosity and understanding.

Over generations, humans continued observing nature with increasing attention. Some individuals began noticing that certain events followed consistent patterns that did not depend on the stories told about them. The sun rose every morning whether people believed it was a god or not. Seeds grew into plants regardless of which myth explained agriculture.

Slowly, a subtle shift began to take place.

Observation started gaining more importance than imagination.

The stories remained, but a new kind of thinking was quietly emerging—the idea that nature might follow rules independent of human narratives.

That realization would eventually transform human curiosity into philosophy and later into science.

1.2 — Nature as Living Presence

For early humans, nature did not appear as a collection of objects and forces. It felt alive.

The wind moved through trees with a voice of its own. Rivers flowed with strength and rhythm. The sun rose every morning with reliability, bringing warmth and light. Storms arrived suddenly, shaking the sky with thunder and lightning. These experiences were powerful and immediate, and they shaped how humans perceived the world around them.

Without scientific tools to measure or analyze these phenomena, the human mind interpreted them through the lens of life and intention. If something moved, acted, or changed the environment, it seemed natural to assume that it possessed some form of agency.

Thunder, therefore, was not merely a sound in the sky. It became the expression of a powerful being. The sun was not simply a ball of burning gas; it was imagined as a traveler crossing the heavens. Rivers were not just flowing water but living entities that nourished the land and demanded respect.

Many ancient cultures reflected this way of thinking. Mountains, forests, oceans, and celestial bodies were often given personalities and stories. These narratives allowed people to form a relationship with the natural world. If the river was a living presence, then it deserved care and caution. If the storm carried intention, then it required respect or appeasement.

In this worldview, humans did not stand apart from nature; they were part of a living environment filled with forces that acted and responded. Fear and reverence often went hand in hand. A forest could provide shelter and food, yet it could also hide danger. Fire could warm and protect, but it could also destroy.

Seeing nature as alive helped early humans make emotional sense of such contradictions. It offered explanations for events that otherwise seemed unpredictable or frightening. By attributing personality or intention to natural forces, humans could create stories that explained why things happened.

These interpretations were not merely imaginative; they were deeply connected to survival and social structure. Shared stories about nature created common understanding within a community. They helped guide behavior—when to plant crops, when to prepare for storms, or how to respect resources that sustained life.

At the same time, this view carried limitations. When natural events were explained through living personalities or divine moods, the explanations often stopped at the story itself. The deeper mechanisms of nature remained hidden behind symbolic meaning.

Yet even within these symbolic interpretations, something important was happening. Humans were recognizing that nature was not random chaos. It possessed patterns, rhythms, and recurring behaviors.

That recognition would later encourage thinkers to ask a new kind of question: if nature behaves in consistent ways, perhaps it follows rules that can be understood without relying solely on stories.

1.3 — Stories as the First Knowledge System

Before books, schools, or written language, knowledge had to travel from one generation to the next in a different way. Humans used stories.

Storytelling was not merely entertainment around a fire. It was the earliest method humans developed to preserve and transmit understanding about the world. In societies without writing, memory was the library, and stories were the structure that allowed knowledge to survive across generations.

Through stories, communities remembered where water could be found during dry seasons. They remembered which plants were safe to eat and which ones caused illness. They remembered the movement of animals during migration and the arrival of changing seasons. These pieces of knowledge were often woven into narratives involving heroes, spirits, or divine forces so that they would be easier to remember and repeat.

A story could travel far and survive long because it engaged the imagination and emotions of listeners. When a child heard the story of a river spirit that protected the valley, the lesson was clear: respect the river, because life depends on it. When a tale described a great storm as the anger of sky gods, the story carried a warning about the power of nature.

In this way, stories functioned as the first knowledge system. They organized observations about the environment into meaningful structures that people could share and recall. The stories did not separate facts from imagination the way modern science does, but they served a practical purpose. They guided behavior and helped communities survive in complex and sometimes dangerous environments.

Stories also created a sense of identity within groups. When people shared the same myths and legends, they shared a common interpretation of the universe. These narratives explained not only natural events but also the origin of tribes, the meaning of life, and the relationship between humans and the world around them.

Over time, certain stories began to reflect recurring natural patterns. Many ancient cultures developed myths that described the cycles of the sun, the moon, and the seasons. While these stories used symbolic language, they often preserved careful observations of nature that had accumulated over generations.

For example, stories about the return of certain stars might signal the beginning of planting season. Legends about changing seasons might encode knowledge about weather patterns or animal migration. Even when wrapped in mythological language, these narratives carried practical insights.

However, stories also had their limitations. They could explain events symbolically but could not always reveal the underlying mechanisms behind them. A myth might describe the sun as a divine traveler crossing the sky, but it could not measure the motion of planets or calculate the forces involved.

Still, the importance of storytelling should not be underestimated. Stories kept curiosity alive. They encouraged people to look at the sky, the land, and the cycles of life with attention and wonder.

In that sense, storytelling was not the opposite of science. It was the first step toward it.

Within every story lived an observation, and within every observation lived a question waiting to be asked.

1.4 — The Limits of Story

Stories helped humans survive and understand their surroundings, but they also carried an invisible boundary.

A story can describe an event and give it meaning, but it cannot always explain how the event actually happens. A myth about the sun traveling across the sky may provide a beautiful image, yet it cannot answer deeper questions about motion, distance, or energy. The narrative stops at symbolism.

For a long time, this limitation was not obvious. The stories seemed complete because they satisfied emotional and cultural needs. They provided comfort, identity, and explanations that felt meaningful within the community.

However, the natural world continued to behave in consistent ways.

The sun rose every morning whether people honored it as a deity or not. Seasons arrived in a regular cycle. Rivers flowed downhill regardless of the stories told about them. Fire burned the same way every time someone placed dry wood into it.

Gradually, attentive observers began noticing that nature followed patterns independent of the stories used to explain it.

This realization was subtle but powerful. If natural events occurred regularly and predictably, perhaps they were governed by rules rather than personalities or moods of invisible beings.

The human mind began to feel the gap between the story and the pattern.

For example, if thunder was the voice of a powerful being in the sky, why did storms follow seasonal patterns? Why did lightning strike certain places more often than others? If illness was caused by supernatural influence, why did some plants consistently heal wounds while others did not?

Questions like these slowly pushed human thinking beyond narrative explanations.

People started comparing events, observing carefully, and remembering what happened under different conditions. Instead of accepting a single story as the final explanation, some individuals began searching for underlying order.

This shift did not happen overnight. Stories continued to exist and remained deeply rooted in culture. But alongside them, a new approach to understanding the world was emerging—one that relied less on symbolism and more on observation and reasoning.

The limitations of storytelling were not failures. They were simply the natural boundary of the first stage of human understanding.

And once humans reached that boundary, curiosity pushed them to cross it.

Beyond that boundary lay the beginning of philosophy.

1.5 — When Stories Begin to Transform

Stories did not disappear when humans began to notice their limitations. They slowly started to change.

Within every community there were individuals who looked at the world with a slightly different attitude. They listened to the same stories as everyone else, but something inside them remained restless. The stories explained events, yet the mind kept asking deeper questions.

If thunder was the voice of a powerful being, why did it follow the same patterns during certain seasons? If the sun was a divine traveler crossing the sky, why did its path remain consistent day after day? Why did the moon change shape in a predictable cycle?

These questions did not reject the stories immediately. Instead, they began to look beyond the story.

Curiosity slowly shifted the focus from *who causes an event* to *how the event happens*. That shift may appear small, but it marked a profound transformation in human thinking.

When people began asking how the world works, they started paying closer attention to patterns. The movement of stars was observed more carefully. The behavior of plants, animals, rivers, and seasons became subjects of thoughtful reflection.

Gradually, explanations moved from narrative symbolism toward reasoning.

Instead of saying the world behaved according to the moods of unseen forces, some thinkers proposed that nature might operate according to principles that could be understood by the human mind. The universe might follow an order that existed whether or not people told stories about it.

This was the birth of philosophical thinking.

Philosophers did not yet have laboratories or scientific instruments. They relied on observation, logic, and careful thought. Their questions were bold for their time: What is the world made of? Why does change occur? What causes motion? Is there an underlying order in nature?

These early thinkers were not scientists in the modern sense, but they were taking the first step away from purely mythological explanations. They were searching for rational understanding.

In this way, stories began their transformation.

The same curiosity that once produced myths now began producing questions about the structure of reality itself. The universe was no longer seen only as a stage for symbolic narratives. It was becoming a system that could be observed, reasoned about, and eventually understood.

This transformation did not erase storytelling. Stories continued to inspire imagination and cultural meaning. But alongside them, a new intellectual path had opened—one that would eventually lead humanity toward philosophy, science, and engineering.

The next stage of this journey begins when human curiosity turns fully toward thought itself: the birth of philosophy.

Closing Transition

Stories were humanity's first bridge to the unknown.

They gave shape to fear, comfort to uncertainty, and meaning to the vast forces of nature that early humans could neither control nor fully understand. Through stories, the universe became something people could speak about, remember, and pass to the next generation. In many ways, storytelling was the first step in humanity's long journey of understanding the world.

But stories carried within them something even more powerful than explanation.

They carried curiosity.

Every story about thunder, the sun, or the changing seasons began with someone noticing something unusual or mysterious. A question appeared first, and the story followed as an attempt to answer it. Over time, however, the questions grew stronger than the stories themselves.

Some minds were no longer satisfied with symbolic explanations. They began to look more closely at the patterns hidden inside nature. They noticed that events repeated with remarkable consistency. The sun rose every morning. The moon followed a cycle. Plants grew under certain conditions. Rivers followed predictable paths.

These observations slowly revealed an important idea: nature might not be governed by unpredictable forces or changing moods. Instead, it might operate through consistent principles that the human mind could discover.

Once that realization began to take root, the role of stories started to shift. Stories remained part of culture, but they were no longer the only way to interpret the universe. Curiosity was moving toward a new stage—one where questions would be explored through reasoning rather than narrative alone.

This moment marked a quiet but profound transition in human thought.

The universe was no longer only something to be imagined.

It was something to be understood.

In the next chapter, we explore the forces that pushed humans to ask these deeper questions. Fear of danger, wonder at the beauty of nature, and the constant struggle for survival all played a role in shaping the earliest philosophical thinking about the world.

Chapter 2 — Fear, Wonder, and Survival

Why survival instincts, fear, and curiosity drove early thinking

Long before humans built cities, machines, or scientific laboratories, life was a constant negotiation with nature.

Imagine waking up in a world where survival depended on understanding your surroundings every moment. The forest could provide food, but it could also hide predators. A river could sustain life, but it could also flood and destroy. A clear sky could suddenly transform into a violent storm.

Early humans lived inside this uncertainty every day.

There were no forecasts, no medicine, no secure shelters, and no guarantees about the next meal. Darkness itself was a source of danger. When night arrived, vision faded and unfamiliar sounds filled the air. A rustle in the bushes could mean wind—or a predator waiting in silence.

In such a world, fear was not merely an emotion. It was a signal.

Fear sharpened attention. It forced the mind to notice small details that might otherwise be ignored—the movement of grass, the direction of animal tracks, the changing color of the sky before rain. The ability to recognize patterns quickly could mean the difference between life and death.

Yet fear was not the only force shaping the early human mind.

Alongside fear lived something quieter but equally powerful: wonder.

Even in the struggle for survival, humans could not ignore the beauty and mystery around them. The night sky stretched endlessly above their heads. The stars appeared in repeating patterns. The moon changed shape over time. The sun returned every morning with remarkable reliability.

These sights did more than provide light or warmth—they stirred curiosity.

The human mind began to ask questions.

Why does the sun return every day? Why do storms arrive suddenly and disappear just as quickly? Why do plants grow from tiny seeds placed in the soil? Why do people live, grow old, and eventually die?

Fear made humans cautious observers. Wonder made them thoughtful observers.

Together, these forces shaped the earliest form of thinking about the world.

Survival demanded awareness, but curiosity pushed awareness beyond immediate needs. The mind did not stop at avoiding danger; it began trying to understand the patterns hidden within nature itself.

From this mixture of fear, wonder, and survival instinct, the first seeds of human thought began to grow.

Civilization would later call this thinking philosophy, science, and engineering.

But at the beginning, it was simply the mind trying to make sense of a powerful and mysterious world.

2.1 — Fear as the First Teacher

For early humans, fear was not an occasional emotion. It was a constant companion.

Every day presented situations that could threaten survival. A sudden storm could destroy shelter. A predator could appear silently in tall grass. A poisonous plant could cause illness. Even unfamiliar sounds in the darkness could signal danger.

In such an environment, the ability to detect risk quickly became one of the most important skills a human could possess.

Fear, therefore, acted as the first teacher of awareness.

When a person feels fear, the mind becomes alert. Attention sharpens. Small details that might otherwise go unnoticed suddenly become important. Early humans learned to read subtle signals from their surroundings: the direction of animal footprints, the smell of smoke, the movement of birds in the sky, or the sudden silence of a forest.

Each of these observations could provide information about what might happen next.

For example, if birds suddenly flew away from a particular area, it might indicate the presence of a predator. If clouds gathered in a certain pattern, rain might soon follow. If animals avoided a particular plant, that plant might be dangerous to eat.

Fear encouraged humans to notice such patterns and remember them.

Over time, these repeated observations slowly formed a practical understanding of the environment. People learned which places were safe to sleep, where water could be found during dry seasons, and which animals were likely to attack.

This learning process was not written in books or taught through formal lessons. It happened through experience, memory, and shared stories within the group.

A child growing up in such a world quickly learned to pay attention. Elders might point out animal tracks, warn about certain plants, or explain signs of approaching storms. Each lesson was connected to survival.

Fear made these lessons unforgettable.

It also pushed humans to search for explanations. If something dangerous happened—such as a sudden illness or an unexpected attack—people naturally wanted to understand why it occurred so they could avoid it in the future.

In this way, fear encouraged both observation and reflection.

The mind began looking not only at what happened but also at what might cause it.

Although early explanations often took the form of stories or myths, the underlying process had already begun: humans were learning to read patterns in nature.

Fear had opened the door to awareness.

And awareness would eventually lead to deeper questioning about how the world truly works.

2.2 — Survival and the Birth of Awareness

Fear helped humans stay alert, but survival required more than momentary reactions. To live successfully in a harsh and unpredictable environment, early humans had to begin recognizing relationships between events.

At first, these realizations were simple.

If fire burned the skin, it had to be handled carefully. If animals gathered near a water source, that location became important for hunting. If certain fruits caused illness, they had to be avoided. These experiences gradually taught humans that actions and consequences were connected.

This was the earliest recognition of cause and effect.

People began noticing that the world behaved in consistent ways. A spark could start a fire. A seed placed in soil could grow into a plant. Rainfall changed the availability of food. Cold seasons demanded preparation.

None of these observations required scientific instruments. They emerged naturally from living closely with the environment. Over time, humans started developing a deeper sense of awareness about the rhythms of nature.

This awareness went beyond immediate survival. It allowed people to anticipate events before they occurred.

For example, the movement of animals might indicate the arrival of a new season. The appearance of certain stars in the night sky might signal changes in weather patterns. The color of clouds could warn of approaching storms.

Gradually, humans were no longer reacting blindly to events. They were learning to interpret signals from the environment.

This ability represented an important shift in human thinking. The mind was beginning to organize observations into patterns that could guide future actions.

Such awareness improved survival dramatically. Groups that understood the behavior of their surroundings could prepare for difficult conditions, find resources more efficiently, and protect themselves from danger.

Awareness also strengthened cooperation within communities. Knowledge gained by one person could be shared with others, helping the entire group adapt to the challenges of nature.

Through this process, the human mind started building its first mental map of the world.

Although the explanations were still mixed with stories and beliefs, the foundation of something new had appeared: the idea that nature followed recognizable patterns that could be observed, remembered, and understood.

Survival had quietly given birth to awareness.

And awareness would soon lead the human mind toward deeper curiosity about the principles guiding the universe.

2.3 — Wonder: The Quiet Companion of Fear

If fear kept early humans alert to danger, wonder kept their minds open to the beauty and mystery of the world.

The same person who feared a sudden storm could also stand in silent amazement under a clear night sky. The stars appeared in patterns that repeated night after night. The moon slowly changed its shape across the month. The sun returned every morning with dependable certainty.

These experiences stirred something deeper than survival.

They created wonder.

Unlike fear, which demanded immediate action, wonder invited quiet attention. It encouraged people to look longer, to observe more carefully, and to think about what they were seeing. The universe was not only dangerous—it was also astonishing.

A river flowing endlessly across the land, a seed turning into a plant, the changing colors of the sky at sunrise and sunset—these were not just events necessary for survival. They were experiences that made humans pause and reflect.

This emotional response played an important role in the development of human thinking.

Fear taught humans to watch their environment carefully to avoid danger. Wonder encouraged them to keep watching even when there was no immediate threat. The mind began observing nature not only for protection but also out of fascination.

Gradually, humans started recognizing the rhythms hidden within the world around them.

The seasons followed one another in a repeating cycle. The moon completed its phases in a predictable pattern. Certain stars appeared in the sky at particular times of the year. Plants grew, animals migrated, and weather patterns shifted in ways that seemed connected.

These repeating rhythms invited deeper reflection.

Why does the sun rise and set every day? Why does the moon change shape over time? Why do seasons return in the same order year after year?

Such questions were not driven by fear. They came from the quiet curiosity created by wonder.

This curiosity slowly expanded human awareness beyond immediate survival. Instead of focusing only on threats and resources, the mind began exploring the larger structure of the world itself.

In this way, wonder became a powerful companion to fear.

Fear sharpened observation when danger was near.

Wonder kept observation alive even when everything seemed calm.

Together, they encouraged humans to pay attention to the universe with unusual depth. Over time, that attention would transform simple curiosity into deeper thinking about the nature of reality.

And from that thinking, philosophy would eventually emerge.

2.4 — Curiosity as an Evolutionary Advantage

Fear helped humans stay alive. Wonder helped them look at the world more carefully.

But curiosity did something even more powerful—it pushed humans to experiment with the world.

Curiosity is the impulse that asks, *What will happen if I try this?* It turns observation into action. Early humans were not only watching nature; they were slowly learning to interact with it in new ways.

Someone discovered that striking stones together could create sparks. Someone else noticed that certain dry materials could catch fire easily. At some point, a curious mind placed seeds intentionally into the soil and waited to see what would happen.

These small experiments changed human life forever.

Groups that explored their surroundings and tried new ideas gradually gained advantages over those who relied only on tradition. A tribe that learned to control fire could cook food, stay warm in colder climates, and protect itself from predators. A community that understood how plants grew could begin early forms of agriculture.

Curiosity therefore became more than a personal trait—it became an evolutionary advantage.

Humans who asked questions and tested ideas were better able to adapt to changing environments. They learned new ways to hunt, build shelters, preserve food, and navigate unfamiliar landscapes.

Over generations, curiosity slowly transformed human life.

Instead of reacting to nature alone, humans began shaping parts of their environment. Tools were invented. Techniques improved. Knowledge accumulated and spread through stories and shared experiences.

Each discovery built upon previous observations. Fire led to cooking and metallurgy. Stone tools evolved into more refined instruments. Understanding animal behavior improved hunting strategies.

Although these developments were still far from modern science, the underlying process had begun: humans were learning that knowledge could change the way they lived.

Curiosity created a bridge between awareness and innovation.

The mind was no longer satisfied with simply observing patterns. It wanted to explore those patterns, test them, and sometimes even use them.

This shift marked an important moment in human evolution. Curiosity transformed survival from a passive struggle into an active search for understanding.

And once the human mind began experimenting with nature, the path toward systematic thinking became inevitable.

The next step would be the realization that these experiments and observations could lead not only to survival skills, but also to deeper insights about how the universe itself works.

2.5 — The Mind That Refuses to Stop Asking

As fear sharpened awareness and curiosity encouraged experimentation, something deeper began forming inside the human mind—a persistent urge to keep asking questions.

At first, questions were directly tied to survival. People wanted to know where food could be found, how to avoid predators, and when seasons would change. These questions helped communities live longer and safer lives.

But the human mind did not stop there.

Once people began noticing patterns in nature, new questions naturally followed. If certain plants grew only during particular seasons, what caused that cycle? If stars appeared in different positions throughout the year, what determined their movement? If fire could transform wood into heat and light, what was happening within the flames?

These questions were no longer about immediate survival.

They were about understanding the world itself.

Curiosity had evolved from a practical tool into a deeper intellectual force. Humans were beginning to wonder not only how to use nature, but also why nature behaved the way it did.

This change marked an important moment in the development of human thinking.

When a person asks how to survive, the answer may be simple and practical. But when a person asks why something happens, the mind must explore ideas that go beyond immediate experience. It must observe more carefully, compare different situations, and think about relationships between events.

Gradually, this questioning mindset began shaping the way humans approached the world.

Some individuals became known for their thoughtful observations and reflections. They spent time contemplating the behavior of nature, the movement of the heavens, the nature of matter, and the meaning of life itself. Their role in society was not only to help the group survive, but also to explore deeper understanding.

These individuals became the earliest philosophers.

They did not yet have scientific instruments or laboratories, but they possessed something equally important: the willingness to question what others accepted without hesitation.

The human mind had discovered a remarkable ability—it could look at the universe and ask questions that had never been asked before.

And once a mind begins asking such questions, it rarely stops.

From this persistent curiosity would emerge philosophy, and from philosophy would eventually grow science and engineering.

The same instinct that once helped humans survive in forests and caves had now become a force capable of exploring the very structure of the universe.

Closing Transition

Fear taught humans to watch carefully. Wonder taught them to keep watching even when there was no danger. Curiosity taught them to ask questions about what they were seeing.

Together, these three forces quietly shaped the human mind.

At first, they helped people survive in a difficult and unpredictable world. But over time they did something even more important—they pushed humans beyond immediate survival and into deeper reflection about the nature of reality.

Once the mind begins asking *why*, the journey of understanding truly begins.

Early humans had already learned to observe patterns in nature: the regular movement of the sun, the changing phases of the moon, the arrival of seasons, and the behavior of plants and animals. These patterns hinted that the universe might not be random or unpredictable after all.

Instead, it might follow an underlying order.

Recognizing this possibility was a turning point in human thought. The universe was no longer only something to fear or admire. It became something that could be thought about carefully.

Some individuals began stepping back from the everyday struggle for survival to reflect more deeply on these patterns. They asked questions not only about how things happen but about the nature of existence itself.

What is the world made of? Why do things change? What causes motion? Is there a hidden order behind everything we observe?

These questions marked the beginning of a new stage in human thinking.

The storyteller was slowly becoming a thinker.

Myths and legends continued to exist, but alongside them emerged a new way of approaching the universe—through reasoning, reflection, and careful questioning. This approach would eventually shape philosophy, the first disciplined attempt to understand reality through the power of thought.

In the next chapter, we step into that remarkable moment in human history when curiosity moved beyond survival and storytelling, and the human mind began consciously searching for the deeper principles that govern the universe.

The age of philosophy had begun.

Chapter 3 — When Stories Became Questions

The moment humans began asking why instead of only telling stories

For generations, the stories had remained the same.

People gathered around fires at night, just as their ancestors had done before them. The elders spoke of the sky, the storms, the sun, and the moon. They explained thunder as the voice of powerful beings and the changing seasons as part of a divine rhythm guiding life on Earth.

Children listened carefully.

These stories were familiar and comforting. They connected people to their past and helped them understand their place in the world. In a universe that often appeared mysterious and unpredictable, stories gave structure and meaning.

But one evening, something slightly different happened.

A young listener, hearing the same explanation about thunder that had been told many times before, paused for a moment and asked a simple question.

"Why does it happen like that?"

The question did not reject the story. It simply asked for something deeper.

The elders might have answered with another story, adding details or symbols that made the explanation richer. Yet the moment had already changed something important. The mind that asked the question had moved one step beyond accepting tradition.

It had begun to search for understanding.

That small shift—from listening to asking—marked a quiet revolution in human thinking.

For the first time, the human mind was no longer satisfied with explanations that simply gave meaning. It wanted to know how things actually worked.

Why does lightning strike the ground?

Why do the stars move across the sky?

Why does the moon change shape over time?

These questions did not arise from survival alone. They emerged from a deeper curiosity about the structure of the world itself.

The person asking such questions was no longer only a storyteller or a listener.

He or she was becoming something new—a thinker.

This transformation did not happen suddenly or in one place. It appeared slowly across different cultures and generations. Somewhere in every society, there were individuals who looked at the world and felt compelled to ask questions that went beyond tradition.

And once the human mind began asking *why*, the universe itself started to look different.

Stories still existed, but they were no longer the final answer.

They had become the beginning of a question.

3.1 — The First "Why"

For a long time, humans accepted the stories they inherited without questioning them. These stories had been told by elders, repeated through generations, and woven into the identity of the community. They explained why storms came, why the sun rose, and why life followed certain rhythms.

But the human mind has a peculiar habit—it does not remain satisfied forever.

At some point, someone begins to ask a different kind of question.

Instead of asking *what happens*, the mind asks why it happens.

This shift may appear small, but it represents one of the most important transformations in human thinking.

The word *why* forces the mind to look deeper than tradition. It asks whether the explanation behind an event truly matches what we observe in nature. It invites the thinker to examine patterns, compare experiences, and search for underlying causes.

Consider the example of thunder.

For many generations, a story might describe thunder as the voice of powerful beings in the sky. The explanation carries meaning and emotional weight. But when someone asks *why thunder follows lightning* or *why storms appear more often in certain seasons*, the story alone may not provide a satisfying answer.

The question pushes the mind to observe more carefully.

Perhaps lightning always appears before the sound of thunder. Perhaps storms begin when certain clouds gather in the sky. Slowly, attention shifts from the symbolic story to the pattern of events themselves.

This is the moment when curiosity begins separating imagination from investigation.

The act of asking *why* also encourages comparison. A thinker may notice that the same natural event occurs in different places, even among communities with different stories about it. If lightning strikes in many regions of the world regardless of local myths, then the explanation might lie somewhere deeper than the stories themselves.

Such realizations gradually change the way humans approach the universe.

Stories are no longer the final answer. They become starting points—ways to frame the question rather than conclude it.

The human mind begins to recognize that the world might follow principles that exist independently of the narratives people create.

In this way, the simple word why opens a door to a new way of thinking.

It transforms the listener into an observer and the observer into a seeker of understanding.

And once that door opens, it rarely closes again.

3.2 — Pattern Over Narrative

As humans continued observing the world around them, something subtle began to change in the way they interpreted events.

Stories had always provided explanations. They gave meaning to the movement of the sun, the arrival of storms, and the changing seasons. These narratives helped communities understand the world emotionally and culturally.

But nature itself kept repeating the same patterns.

The sun rose every morning and set every evening with remarkable consistency. The moon changed shape in a cycle that repeated month after month. The seasons followed one another in a reliable order—warm months giving way to colder ones, and then returning again.

These patterns were too regular to ignore.

Gradually, attentive observers began noticing that the behavior of nature did not depend on the stories told about it. The sun rose whether it was believed to be a divine traveler or not. Storms followed weather conditions rather than human rituals. Rivers continued flowing regardless of the myths associated with them.

This realization introduced a new way of thinking.

Instead of focusing on the meaning behind an event, the human mind began focusing on the pattern of the event itself.

For example, early observers of the night sky noticed that certain stars appeared at the same times each year. These observations helped communities predict seasonal changes and prepare for planting or migration. The stars were no longer only symbols in stories—they became markers of natural cycles.

Similarly, the repeated behavior of plants and animals revealed rhythms that people could depend on. Seeds sprouted when placed in soil under the right conditions. Animals migrated along predictable routes. Rivers rose and fell according to seasonal rains.

Each pattern provided evidence that nature operated according to a certain order.

The mind slowly began trusting these patterns more than the stories used to describe them.

This shift did not immediately eliminate myths and legends. Stories remained deeply embedded in culture. However, alongside those stories grew a new appreciation for observation and consistency.

People started realizing that understanding patterns could help them predict events more accurately than relying solely on narrative explanations.

This insight represented a critical step in the development of human knowledge.

The universe was beginning to look less like a collection of unpredictable events driven by personalities or moods, and more like a system governed by repeating behaviors that could be observed and studied.

Once the human mind recognized the importance of patterns, the next step became inevitable.

If patterns existed, perhaps there were deeper principles behind them.

And discovering those principles would require a new kind of thinking—one based not on stories, but on reasoning.

3.3 — The Birth of Rational Thought

As patterns in nature became clearer, the human mind began searching for explanations that went beyond storytelling. The question was no longer simply *what happens* or even *when it happens*, but something deeper: *What principle might explain it?*

This marked the beginning of rational thought.

Rational thinking does not reject imagination entirely, but it asks the mind to move beyond symbolic explanations and look for reasons grounded in observation and logic. Instead of saying that thunder occurs because a powerful being in the sky is angry, a rational thinker begins asking: *What physical event might produce that sound?*

Such thinking was revolutionary for its time.

Across different civilizations, individuals began stepping away from mythological explanations and exploring nature through careful reasoning. In ancient Greece, thinkers like Thales and Anaximander wondered what the universe might be made of and whether a basic substance or principle underlay all things. In India and China, philosophers examined the nature of matter, change, and harmony in the universe. These early thinkers were not performing modern experiments, but they were attempting something new: explaining the world through ideas that followed logical consistency rather than narrative tradition.

Their questions were bold.

If everything changes, what causes change? If objects move, what makes them move? Is the universe made of one fundamental substance, or many? Does nature follow rules that remain constant?

By asking such questions, these thinkers introduced the concept that the universe might be governed by principles that can be discovered through reasoning.

This approach slowly reshaped the human understanding of reality.

Nature began to appear less like a stage controlled by invisible personalities and more like a system operating according to underlying order. Instead of focusing on the intentions of supernatural forces, rational thought focused on the structure and behavior of the world itself.

This change did not happen overnight. Stories and myths continued to play important roles in culture and tradition. But alongside them, a new intellectual path was forming—one that valued observation, comparison, and logical explanation.

The birth of rational thought therefore represents a turning point in the human journey of understanding.

For the first time, the mind was not only asking *why* but also attempting to construct explanations that could apply universally, regardless of who observed the event or what story was told about it.

From this shift would emerge the foundations of philosophy and, eventually, the scientific way of thinking that continues to guide human exploration today.

3.4 — From Meaning to Mechanism

The earliest stories about nature focused mainly on meaning.

People wanted to understand what an event meant for their lives. A storm might be interpreted as a warning or punishment. A successful harvest might be seen as a blessing. Natural events were often explained in ways that connected them to human emotions, values, or relationships with unseen forces.

These explanations were meaningful because they helped people make sense of their place in the world.

However, as curiosity grew and observations accumulated, a new kind of question began to appear. Instead of asking *what an event meant*, some thinkers began asking *how the event actually happens*.

This shift moved human thinking from meaning to mechanism.

A story might say that thunder represents the anger of a sky deity. But a curious observer might notice that thunder always follows a flash of lightning. That observation invites a different type of explanation: perhaps the flash and the sound are connected through some physical process.

Similarly, the rising and setting of the sun had long been described through myths of celestial journeys. Yet careful observers noticed that the sun followed a regular path across the sky, and that this path changed gradually throughout the year. This suggested that the movement might follow a predictable pattern rather than a symbolic story.

When people began focusing on mechanisms, they started paying attention to relationships between events.

Fire became more than a mysterious force; it became something that could be produced under certain conditions. A seed placed in soil did not grow because of a story—it grew because of natural processes that could be observed repeatedly. Motion, heat, light, and life itself began to appear as parts of a system rather than isolated mysteries.

This shift had profound consequences.

Understanding mechanism allows humans to predict and influence events. If people understand what causes fire, they can learn how to control it. If they understand how plants grow, they can cultivate crops intentionally. Knowledge of mechanisms transforms curiosity into practical capability.

Gradually, the focus on mechanism began shaping the way humans approached knowledge itself. Instead of accepting explanations that satisfied emotional meaning alone, thinkers started searching for explanations that could describe how nature works consistently.

This way of thinking laid the foundation for future scientific inquiry.

It encouraged people to observe carefully, compare events, and question explanations that did not match what they saw in the natural world. Mechanisms became more valuable than narratives when it came to understanding the structure of reality.

Yet this transformation did not erase meaning from human life. Stories, culture, and symbolism continued to exist alongside emerging rational explanations. The difference was that humans were now learning to separate cultural meaning from natural mechanism.

And once that distinction appeared, the path toward systematic investigation of nature became clearer.

Understanding how things work would eventually lead to the development of experiments, measurements, and scientific methods—tools that would allow humanity to explore the universe with unprecedented depth.

3.5 — Curiosity Becomes a Method

Once humans began asking *why* and looking for mechanisms behind events, curiosity started changing its character.

At first, curiosity was spontaneous. Someone noticed a strange pattern in the sky or wondered why a plant grew better in one place than another. The question might arise from simple observation or personal interest.

But as more people began asking similar questions, a new habit slowly developed: questions were no longer isolated—they began leading to further investigation.

People started comparing their observations with others. They discussed ideas, debated explanations, and refined their thinking. A question asked by one person could inspire another person to look more carefully at the same phenomenon.

Over time, curiosity began forming a process.

Instead of accepting the first explanation that came to mind, thinkers began testing ideas against observation. If someone believed that a certain plant healed wounds, others might try using that plant in similar situations to see if the effect repeated. If someone noticed that stars appeared in certain positions during particular seasons, observers in different places might confirm the same pattern.

Gradually, questioning turned into a form of investigation.

The human mind was learning that knowledge becomes stronger when ideas are examined carefully and compared with experience. A single observation could inspire a story, but repeated observations could reveal a pattern.

This process was still informal compared to modern science, but it marked an important shift. Curiosity was no longer only about asking questions—it was also about finding reliable ways to answer them.

In many ancient cultures, this transition produced thinkers who dedicated their lives to exploring nature, mathematics, and philosophy. They observed the sky, studied numbers, reflected on motion and change, and tried to build explanations that could apply universally.

These individuals were not yet scientists in the modern sense. They did not always perform controlled experiments or use precise instruments. But they were developing something equally important: a method of thinking that valued reasoning, consistency, and observation.

This method gradually transformed curiosity into a disciplined search for understanding.

Ideas were no longer judged only by tradition or authority. They were increasingly judged by whether they matched the behavior of the world itself.

In this way, curiosity became more than an instinct—it became a guiding principle for exploring reality.

The storyteller had become a questioner.

The questioner was becoming a thinker.

And from this transformation, the first systematic philosophies about the universe began to emerge.

Closing Transition

Stories helped humanity begin its journey of understanding, but questions changed the direction of that journey.

Once people began asking *why*, the universe could no longer remain a place explained only through myth and imagination. Curiosity pushed the mind to look deeper, to compare observations, and to search for patterns that existed beyond the stories told around the fire.

Gradually, a new kind of thinker began to appear in human societies.

These individuals were not satisfied with repeating traditional explanations. They spent time reflecting on nature, discussing ideas with others, and exploring the hidden order behind everyday experiences. They looked at the sky, the movement of water, the behavior of fire, and the growth of plants with a different perspective.

Instead of asking only *what does this mean for us*, they began asking *what principle might explain this behavior*.

This change marked the birth of philosophy.

Philosophy was the first systematic attempt to understand the universe using reasoning and reflection. Philosophers tried to identify the basic elements of reality, explain the causes of change, and discover whether nature followed universal laws.

Although they did not yet have the tools of modern science, they had something equally important: the courage to question tradition and the patience to think deeply about the world.

From these early philosophical explorations would eventually emerge the foundations of scientific thought. Careful observation would lead to measurement. Logical reasoning would lead to experimentation. Curiosity would lead to discovery.

The storyteller had become a thinker.

And the thinker was beginning to explore the universe not through myth, but through the power of thought.

In the next part of our journey, we enter the age of philosophy—when human reasoning began shaping the first organized attempts to explain the nature of reality.

Reflective Interlude

From Stories to Questions

We have now completed the first stage of our journey.

In the beginning, humans looked at the universe with a sense of fear and wonder. The world was powerful, unpredictable, and mysterious. Storms shook the sky, fire burned with dangerous beauty, rivers flowed endlessly, and the stars appeared each night like silent witnesses to human life.

Without tools, measurements, or formal knowledge, the human mind responded in the only way it could—through stories.

Stories helped people cope with uncertainty. They gave meaning to natural events and created a shared understanding within communities. Through myths and legends, early humans explained why the sun moved across the sky, why seasons changed, and why life itself followed a cycle of birth and death.

These stories were not mistakes.

They were the first language humans used to speak about the universe.

But hidden inside every story was something more powerful than the story itself.

There was a question.

As humans continued observing the world, curiosity slowly grew stronger than the narratives that once satisfied the mind. People began noticing patterns that did not depend on the stories told about them. The sun rose every morning. The moon followed a repeating cycle.

Plants grew under certain conditions, and storms arrived in recognizable ways.

Once these patterns became clear, a new kind of thinking began to appear.

The human mind started asking *why*.

Why does the sun follow the same path across the sky? Why do seasons repeat year after year? Why does fire behave the same way every time it burns?

The storyteller was slowly becoming a questioner.

This shift may appear simple, but it changed the course of human civilization. The moment a question replaces a story as the center of attention, the search for deeper understanding begins.

The human mind had discovered something extraordinary—it could reflect on the universe itself.

That discovery opened the door to the next stage of our journey.

Instead of only telling stories about the world, humans would begin thinking about the principles that might govern it. They would ask what the universe is made of, what causes motion and change, and whether there is an underlying order behind everything we observe.

This stage is called **philosophy**.

In the next part of this book, we will explore how human reasoning began transforming curiosity into structured thought. The earliest philosophers did not yet possess laboratories or modern instruments, but they possessed something equally important: the courage to question the universe and the patience to think deeply about its nature.

The journey from stories to questions has brought us to the threshold of philosophy.

Now the human mind is ready to explore the universe through thought itself.

PART II — PHILOSOPHY

"When humans began asking why instead of simply accepting what is, thinking itself became a new form of exploration."

Chapter 4 — The Birth of Thought

How early philosophers tried to explain the world through reasoning

At some point in human history, a quiet transformation began to take place.

People were still telling stories about the universe. Myths about the sun, storms, rivers, and stars continued to shape the cultural life of communities. But somewhere among those communities, a few individuals began looking at the same world with a slightly different attitude.

Imagine a person sitting near a flowing river.

The water moves steadily, following its path around stones and through the land. The sun slowly travels across the sky, casting changing shadows as the day progresses. At night, the stars appear in patterns that repeat again and again across the seasons.

The observer watches these events carefully.

For generations, stories had explained them. The river might have been described as the presence of a spirit. The sun might have been imagined as a divine traveler crossing the heavens. The stars might have been seen as lights placed in the sky by powerful beings.

But this observer is not fully satisfied with the story alone.

Instead of asking what the event symbolizes, the mind begins asking something more fundamental.

What actually causes the river to flow? Why does the sun move across the sky in the same pattern every day? Why do the stars appear in predictable positions throughout the year?

These questions do not reject tradition or cultural meaning. They simply look beyond it.

For the first time, the human mind is trying to understand the universe through reasoning.

Reasoning means searching for explanations that make sense based on observation and logical thinking. Instead of relying only on inherited stories, the thinker begins exploring ideas about how the world might actually be structured.

This moment marks the birth of philosophy.

The early philosophers were not scientists in the modern sense. They did not possess telescopes, microscopes, or laboratories. Yet they possessed something that would become equally powerful: the determination to think carefully about reality itself.

They asked questions that had rarely been asked before.

What is the universe made of? Why does change occur? What causes motion? Is there an underlying order behind everything we observe?

These questions may appear abstract, but they represent one of the most important steps in human intellectual history.

The storyteller had become a thinker.

And once humans began exploring the universe through thought, a new path of understanding opened—one that would eventually lead to science, mathematics, and engineering.

4.1 — The Courage to Question Tradition

For thousands of years, stories had guided human understanding of the world. They were respected, repeated, and rarely questioned. Stories carried the wisdom of ancestors, and to challenge them could appear disrespectful or even dangerous.

In many early societies, tradition held great authority. What elders said was considered reliable because it had been preserved across generations. Stories about the sun, storms, rivers, and the origin of life were deeply woven into culture and belief.

Yet the birth of philosophy required something unusual: the courage to question tradition.

Early philosophers were not simply rebellious individuals rejecting culture. Most of them grew up listening to the same stories as everyone else. They respected the wisdom of their communities. But curiosity pushed them to look beyond the explanations that had always been accepted.

When a philosopher asked questions such as *What is the world made of?* or *What causes change in nature?*, the answer could not be found within familiar myths alone. The thinker had to rely on observation, reflection, and logical reasoning.

This required intellectual courage.

Imagine living in a society where everyone believes that lightning occurs because powerful beings in the sky are expressing anger. To suggest that lightning might be the result of natural processes rather than divine moods could challenge deeply held beliefs.

Early thinkers had to walk carefully between respect for tradition and the desire for deeper understanding.

Some philosophers began suggesting that natural events might have explanations rooted in the structure of the world itself. They proposed that the universe might be governed by principles that exist independently of human stories.

This idea was revolutionary.

It suggested that the world could be understood through observation and reasoning rather than through myth alone. Instead of asking what a storm meant symbolically, a philosopher might ask what physical conditions caused the storm to form.

Across different civilizations, thinkers began exploring such questions.

In ancient Greece, philosophers wondered about the fundamental substance underlying all matter. In India and China, scholars explored the nature of change, harmony, and balance within the universe. Though their ideas varied, they shared a common impulse: to understand the world through thoughtful inquiry.

Questioning tradition did not immediately replace myths and stories. Cultural narratives continued to play an important role in society. However, a new intellectual space had opened—one where ideas could be examined through reasoning and discussion.

Once this space existed, the human mind gained the freedom to explore deeper questions about reality.

And from that freedom, philosophy began to flourish.

4.2 — The Search for the First Principle

Once early thinkers began questioning traditional explanations, a natural next step followed. If the world behaves in organized ways, perhaps there is something fundamental beneath all the complexity.

This led philosophers to search for what they called a first principle—a basic substance or rule from which everything else might arise.

The idea was simple but powerful.

The world appears filled with countless forms: mountains, rivers, plants, animals, clouds, stars, and people. Yet beneath this diversity, philosophers wondered if there might be a smaller set of underlying elements or principles that create everything we see.

In ancient Greece, some of the earliest philosophers tried to answer this question by proposing simple substances. Thales suggested that water might be the fundamental substance of the universe, since it appeared in many forms—liquid, vapor, and ice—and seemed essential for life. Anaximenes believed air might be the primary element, capable of transforming into other forms through compression and expansion.

These ideas may sound primitive today, but they represented an important shift in thinking. Instead of explaining the world through mythological characters or symbolic narratives, these philosophers were searching for universal explanations that could apply everywhere.

Similar explorations appeared in other civilizations.

In ancient India, philosophers developed theories about fundamental elements such as earth, water, fire, air, and space, believing that all physical forms emerged from combinations of these elements. In China, thinkers proposed systems such as the five phases—wood, fire, earth, metal, and water—to describe patterns of change and interaction in nature.

Although these systems differed in detail, they shared a common goal: identifying the underlying structure of reality.

The search for a first principle was not only about matter. Some philosophers believed that the universe might be governed by an abstract principle rather than a physical substance. They suggested that order, balance, or harmony could be the true foundation of the world.

For example, the Greek philosopher Heraclitus emphasized constant change, proposing that reality is defined by transformation rather than permanence. Others, like Pythagoras, believed that numbers and mathematical relationships might reveal the hidden order of the universe.

These thinkers were attempting something remarkable: explaining the entire cosmos using reasoning.

They did not yet have experimental methods or advanced instruments. Their tools were observation, logic, and imagination disciplined by thought. Even when their specific answers were incomplete or incorrect, the method itself represented a major step forward.

The search for the first principle showed that the human mind was beginning to look for universal patterns beneath the surface of reality.

This search would eventually lead to more refined theories about matter, energy, and the structure of the universe. But at its heart, the effort reflected a simple desire: to understand whether the complexity of the world could be traced back to something fundamental.

And that question remains at the center of human inquiry even today.

4.3 — Order Instead of Chaos

As early philosophers continued reflecting on nature, they began noticing something remarkable.

The universe did not behave like a chaotic collection of random events.

Instead, it appeared to follow consistent patterns and rhythms.

Day followed night with predictable regularity. The seasons returned in a repeating cycle. The movement of stars across the sky followed recognizable paths. Rivers flowed according to the shape of the land, and plants grew under certain conditions again and again.

Such observations suggested that the world might not be governed by unpredictable forces after all.

It might follow order.

This realization was one of the most profound insights in the early development of philosophy. If the universe followed order, then events were not happening purely by accident. There must be principles or relationships guiding the behavior of nature.

Some philosophers began exploring this idea more deeply.

For example, the Greek thinker Heraclitus observed that change seemed to be a constant feature of the world. Rivers flowed continuously, fire transformed materials, and life itself moved through cycles of growth and decay. From these observations, he suggested that reality might be governed by a principle of continuous change.

In contrast, another philosopher, Parmenides, argued that beneath the appearance of change there must exist something stable and unchanging. If everything were constantly changing without any underlying stability, he reasoned, the world could not exist in a coherent form.

These different perspectives revealed an important aspect of philosophical thinking: philosophers were no longer satisfied with surface explanations. They were trying to understand the deeper structure behind the world's behavior.

Across other cultures, similar ideas appeared.

In ancient China, thinkers explored the concept of harmony and balance within nature. They believed that opposing forces interacted in ways that produced stability and change. In India, philosophical traditions examined the principles governing transformation and permanence within the universe.

Although the details varied, these thinkers shared a common insight.

Nature seemed to follow rules.

This idea gradually transformed the way humans viewed the universe. Instead of imagining nature as a stage controlled by unpredictable personalities or divine moods, philosophers began viewing it as a system governed by underlying order.

Once this possibility was recognized, an important question emerged.

If nature follows order, can the human mind discover the principles behind that order?

This question would eventually lead to the development of mathematics, science, and systematic investigation of the natural world.

But at this stage, philosophy had already accomplished something extraordinary.

It had introduced the idea that the universe might be understandable.

4.4 — Thinking as a Tool

The early philosophers did not possess telescopes, microscopes, or laboratories. They did not have instruments capable of measuring the smallest particles or the distant stars. Yet they were attempting to understand the entire universe.

Their primary tool was something much simpler and yet incredibly powerful: thought itself.

Thinking became an instrument for exploring reality.

By carefully observing nature and reflecting on those observations, philosophers began constructing ideas about how the world might work. They discussed their ideas with others, examined contradictions, and tried to build explanations that remained logically consistent.

This process may seem ordinary today, but at that time it represented a new intellectual approach.

Instead of accepting explanations simply because they were traditional, philosophers began asking whether an explanation made sense when examined through reasoning.

For example, if someone claimed that a natural event occurred because of the will of unseen forces, a philosopher might ask: *Does this explanation account for the patterns we observe?* If the event follows the same pattern every time, perhaps there is a more consistent reason behind it.

Reasoning allowed philosophers to explore questions that could not yet be tested directly.

They debated the nature of matter, wondering whether everything might be composed of tiny invisible particles. They discussed whether motion required a continuous force or whether objects might move according to natural tendencies. They explored ideas about infinity, space, and the structure of the cosmos.

These discussions did not always produce correct answers, but they created something even more important: a culture of intellectual inquiry.

Philosophers learned that ideas could be examined, compared, and refined through conversation and reflection. One thinker might propose a theory, another might challenge it, and through this exchange new insights could emerge.

This process helped sharpen the ability of the human mind to analyze complex questions.

In many ways, philosophical reasoning functioned like an early laboratory of ideas. Instead of conducting physical experiments, philosophers conducted experiments of thought. They imagined different possibilities, tested them against observation and logic, and explored their implications.

Over time, this habit of disciplined thinking became a powerful intellectual tradition.

It prepared the human mind for the next stage in the development of knowledge. Once reasoning had clarified questions about the world, people began searching for ways to test those ideas more precisely.

Observation would soon be joined by measurement.

Reasoning would soon be joined by experimentation.

But before science could emerge, the human mind first had to learn how to use thinking itself as a tool.

4.5 — The Seeds of Science

Although early philosophy relied mainly on reasoning and observation, it quietly planted the foundations for what would later become science.

Philosophers were not yet conducting controlled experiments or using mathematical formulas to describe natural laws. But they were developing habits of thinking that would make scientific inquiry possible.

One of the most important of these habits was the belief that the universe could be understood through consistent principles.

If nature followed order rather than random chaos, then careful observation and logical reasoning might reveal the rules behind that order. This idea alone was powerful enough to change the direction of human curiosity.

Philosophers also emphasized the importance of questioning assumptions. Instead of accepting explanations simply because they were traditional or widely believed, they encouraged people to examine ideas critically.

This spirit of questioning became essential for future scientific thinking.

For example, if someone proposed that all matter was composed of a particular substance, philosophers would ask whether the idea explained different observations consistently. If an explanation produced contradictions, it needed to be reconsidered.

Such discussions trained the human mind to evaluate ideas not by authority alone, but by coherence and evidence.

Another important contribution of philosophy was the recognition that knowledge could be shared and refined through dialogue.

Philosophers often debated their ideas with others, testing arguments and exploring alternative explanations. Through these conversations, new insights emerged and older ideas were improved or replaced.

This collaborative process would later become a central feature of scientific communities.

Gradually, philosophers also began recognizing the value of careful observation. Although they did not yet have precise instruments, they paid attention to natural patterns and attempted to describe them systematically.

In some cases, philosophical ideas even hinted at concepts that would later become scientific discoveries. Ancient thinkers speculated about the existence of atoms, the nature of motion, and the structure of the cosmos—ideas that would eventually be explored through experimental science centuries later.

In this way, philosophy prepared the intellectual ground for science.

It introduced the belief that the universe could be understood through reasoning, that ideas could be examined critically, and that knowledge could grow through discussion and observation.

When later thinkers began adding measurement, experimentation, and mathematics to these philosophical foundations, science as we know it began to emerge.

The philosopher's questions would soon become the scientist's investigations.

Closing Transition

With the birth of philosophy, something remarkable had happened to the human mind.

For thousands of years, humans had looked at the universe through stories. These stories gave meaning to storms, stars, rivers, and the cycle of life. They helped communities live together and pass knowledge from one generation to the next.

But once philosophical thinking began, the focus slowly shifted.

The universe was no longer only something to admire, fear, or interpret symbolically. It had become something that could be examined through reasoning.

Philosophers began asking questions that went beyond immediate experience. They wondered what the universe was made of, what causes change and motion, and whether nature followed universal principles that could be discovered by the human mind.

This shift transformed curiosity into structured thought.

Even though early philosophers did not yet possess the tools of modern science, they had already taken the most important step. They believed that the universe was not random chaos and not merely a stage for stories, but a system that could be understood through careful thinking.

From this belief emerged the intellectual courage to explore deeper patterns in nature.

Gradually, philosophical thinking began revealing something surprising: the world around us seemed to follow consistent relationships and rhythms. The movement of the heavens, the behavior of matter, the cycles of life and energy—all hinted at an underlying order.

Understanding that order became the next great challenge for human thought.

If the universe follows patterns, how can we recognize them clearly? If nature operates through principles, how can we describe them accurately?

These questions would lead thinkers toward new tools for understanding reality.

Observation would become more careful. Reasoning would become more systematic. Mathematics would begin describing patterns with precision.

The philosopher had opened the door.

Soon, the scientist would step through it.

In the next chapter, we explore how philosophers began identifying harmony and structure within nature itself, discovering that the universe might be governed by patterns far more elegant than anyone had previously imagined.

Chapter 5 — Order in Nature

Discovering patterns in seasons, motion, life, and the sky

Imagine living in a world where every day brings change, yet over time you begin to notice something remarkable: the changes are not random.

At first, the world appears unpredictable. Storms arrive suddenly. Rivers flood and then retreat. Plants grow, animals migrate, and the sky shifts from light to darkness every day. To someone observing these events for the first time, nature might seem chaotic.

But patience reveals something different.

An attentive observer begins noticing that many of these changes repeat themselves in recognizable patterns. After cold months, warmth slowly returns. After the harvest, the land rests before new growth begins again. The moon appears thin, then full, then disappears before repeating the cycle once more.

These patterns are subtle at first, but once noticed, they become impossible to ignore.

Early humans living close to nature would have experienced these rhythms directly. Their survival depended on recognizing them. Knowing when seasons would change meant knowing when to plant crops or when animals might migrate. Watching the sky helped people anticipate weather and prepare for long journeys.

Over time, observation revealed that nature was not simply a series of isolated events. It behaved more like a vast system of repeating cycles.

The sky itself became one of the clearest examples of this order. The sun followed a path across the sky every day. The moon returned to familiar shapes across the month. Certain stars appeared in specific seasons, guiding travelers and marking the passage of time.

Such regularity suggested something profound.

Perhaps the universe was not governed by random events or unpredictable forces. Perhaps it followed an underlying order.

This realization changed the way humans thought about the world. Instead of seeing nature only as mysterious or symbolic, thinkers began viewing it as something that could be studied and understood.

Patterns became clues.

Each repeating cycle hinted that the universe might operate according to principles that the human mind could discover. The rhythms of seasons, the motion of celestial bodies, and the cycles of life all pointed toward the same idea: beneath the complexity of nature lies an elegant structure waiting to be understood.

Once this possibility appeared, the search for that hidden order became one of the greatest intellectual adventures in human history.

5.1 — The Rhythm of Seasons

Among the first patterns humans recognized in nature was the rhythm of the seasons.

Long before calendars or scientific instruments existed, people living close to the land noticed that the environment changed in repeating cycles. Warm periods were followed by colder months. Rainy seasons arrived after dry ones. Plants grew, produced seeds, and then withered before new growth began again.

These changes were not random.

They returned year after year with surprising regularity.

For early human communities, recognizing these patterns was essential for survival. Knowing when certain plants would grow or when animals would migrate helped people prepare for hunting and gathering. As humans gradually developed agriculture, understanding seasonal rhythms became even more important. Crops had to be planted at the right time and harvested before unfavorable weather arrived.

Careful observation allowed communities to anticipate these changes.

People noticed that certain natural signs often appeared before a shift in season. The length of daylight changed slowly throughout the year. The position of the sun at sunrise and sunset moved slightly along the horizon. Some stars became visible only during particular months.

These observations provided valuable clues.

For example, the appearance of specific stars in the night sky could signal the beginning of planting season. The behavior of birds or animals could indicate approaching weather changes. Over time, humans began connecting these signals with the larger cycle of the year.

Gradually, seasonal patterns became one of the earliest examples of predictable order in nature.

Communities began organizing their activities around these cycles. Festivals celebrated the return of spring or the completion of harvests. Rituals and traditions often reflected important moments in the agricultural calendar. Even myths and stories began incorporating seasonal themes.

Although these cultural expressions explained seasons symbolically, the underlying observations were grounded in careful attention to nature.

The recognition of seasonal rhythms also encouraged people to think more deeply about the forces shaping the world. If seasons followed a consistent pattern, something must be controlling this cycle. Why did warm and cold periods alternate? Why did daylight change across the year?

These questions slowly guided human curiosity toward the sky.

The movement of the sun and stars seemed closely connected to the changing seasons. Observing the heavens therefore became a natural next step in understanding the order of nature.

And so, the sky became humanity's first great laboratory.

5.2 — The Sky as Humanity's First Laboratory

While the land revealed seasonal rhythms, the sky offered something even more precise: predictable motion.

For early observers, the sky was impossible to ignore. It stretched endlessly above them, changing slowly yet consistently. Every day the sun appeared in the east, climbed across the sky, and disappeared in the west. Night followed day without fail.

But careful observers soon noticed something more subtle.

The sun did not rise at exactly the same point on the horizon every day. Across the year its position shifted slightly north and south before returning again. These gradual changes were connected to the seasons that people experienced on Earth.

The moon offered another pattern.

Unlike the sun, it did not remain the same shape each night. Sometimes it appeared as a thin curve, sometimes half-lit, and sometimes as a bright full circle dominating the sky. After reaching fullness, it slowly diminished until it disappeared completely—only to begin the cycle again.

This repeating sequence was one of the earliest natural clocks available to humanity.

Then there were the stars.

At first glance the stars seemed fixed in the sky, forming recognizable patterns that later became constellations. Yet over long periods, observers noticed that these patterns moved slowly across the heavens during the night. More importantly, certain stars appeared only during particular seasons.

For people living without artificial light, the night sky became an extraordinary source of information.

Travelers used the stars to navigate across deserts and oceans. Farmers observed celestial patterns to anticipate seasonal changes.

Entire civilizations began organizing their calendars according to the movements of the sun and moon.

The sky had become humanity's first laboratory.

Unlike experiments performed in modern laboratories, these observations required patience rather than equipment. Generations of observers watched the heavens, recording patterns through memory, stories, and eventually early forms of written records.

Ancient monuments in different parts of the world—such as stone circles, pyramids, and temples—often aligned with the positions of the sun during solstices or equinoxes. These structures suggest that early civilizations were carefully studying celestial motion and using that knowledge to mark important moments in the yearly cycle.

What made the sky particularly powerful for early thinkers was its reliability.

The motions of the sun, moon, and stars appeared remarkably consistent. Unlike storms or animal behavior, celestial patterns repeated with almost perfect regularity.

This reliability inspired an important idea.

If the heavens follow precise patterns, perhaps the universe itself operates according to predictable laws.

The sky therefore played a crucial role in shaping human thought. By observing celestial motion, people began realizing that nature might not be random after all.

Instead, the universe might be governed by an underlying order that could be discovered through careful observation and reasoning.

And once humans began recognizing such order in the heavens, they naturally started looking for similar patterns everywhere else in the world.

5.3 — Motion Everywhere

As early thinkers observed nature more carefully, one feature of the world became impossible to ignore.

Everything seemed to be in motion.

Rivers flowed continuously across the land. Wind moved through forests, bending branches and carrying clouds across the sky. Fire flickered and transformed the materials it touched. Animals traveled across plains and forests in search of food and shelter. Even the heavens themselves appeared to move, with the sun crossing the sky each day and the stars shifting position through the night.

Motion was not an occasional event—it was everywhere.

This observation raised a deep philosophical question: What causes motion?

For people who had grown up with stories explaining natural events, motion might have been interpreted as the action of invisible forces or divine beings. But philosophers began approaching the question differently.

Instead of asking *who causes motion*, they asked *what principle might explain it.*

If rivers flow downhill, perhaps the shape of the land influences the movement of water. If wind moves through the air, perhaps invisible currents exist that push the air in certain directions. If celestial bodies move across the sky in predictable patterns, perhaps there are rules governing their motion.

Such questions revealed that motion might not be random or symbolic—it might follow natural principles.

Different philosophers proposed different ideas.

Some thinkers believed that motion was connected to change. Everything in the world, they argued, was constantly transforming. Fire turns wood into heat and ash. Seeds grow into plants. Day becomes night. From this perspective, motion was simply the visible expression of the universe's ongoing transformation.

Others suggested that motion must have causes that could be traced back to fundamental forces or interactions. If something moves, perhaps something else must push or pull it.

These early reflections may seem simple today, but they represented an important shift in thinking. Motion was no longer just something that happened—it had become a subject of investigation.

Philosophers began asking questions that would later become central to physics:

Why do objects move? What determines the path of moving bodies? Why do celestial bodies follow regular patterns across the sky?

Even though early philosophers did not yet possess the tools to measure motion precisely, their questions opened the door to deeper exploration.

Motion, it seemed, was not merely a feature of isolated events. It was a fundamental characteristic of the universe itself.

Recognizing this would eventually lead thinkers toward the development of mechanics, astronomy, and the mathematical description of motion—fields that would transform humanity's understanding of the cosmos.

But at this stage, the important realization had already been made:

The universe was not static.

It was alive with movement, and that movement hinted at hidden rules waiting to be discovered.

5.4 — Patterns in Life

While the sky revealed cycles of celestial motion and the land revealed the rhythm of seasons, life itself offered another powerful example of order in nature.

Early humans observed the living world closely because their survival depended on it. Plants provided food and medicine. Animals were both prey and predators. The health of the environment directly affected the well-being of the community.

Over time, attentive observers began noticing that life followed recognizable patterns.

Plants grew from seeds placed in soil. They matured, produced new seeds, and eventually withered. The next generation of plants would then grow from those seeds, repeating the cycle again and again.

Animals followed similar rhythms.

Many species migrated during certain times of the year. Others reproduced during particular seasons. Young animals grew into adults and later produced offspring of their own.

Human life followed a comparable pattern.

A child was born, grew into adulthood, aged, and eventually passed away. New children were born, continuing the cycle of life within the community.

These repeating processes suggested that life itself was not random.

It followed cycles.

The recognition of these cycles helped early thinkers understand the interconnectedness of living systems. The growth of plants depended on sunlight, water, and soil. Animals depended on plants or other animals for food. Humans depended on both plants and animals to survive.

Gradually, people began realizing that life was part of a larger system of relationships within nature.

The birth of new plants depended on seeds. Seeds depended on previous plants. Animals depended on plants for nourishment. Humans depended on the stability of these natural processes.

This web of relationships revealed that living systems were governed by patterns just as clearly as the sky or the seasons.

Philosophers began reflecting on these patterns with deeper curiosity.

Why does life follow cycles of birth and death?

What causes growth and transformation in living organisms?

Is there a principle that connects the processes of life with the patterns observed in nature?

Some thinkers believed that life represented a form of harmony within nature. Others proposed that life followed natural processes similar to those governing the physical world.

Although early explanations varied widely, the observation itself was important: life follows order.

Recognizing this order encouraged humans to study living systems more carefully. It also strengthened the belief that nature might operate according to principles that apply across many different phenomena.

If the sky, the seasons, motion, and life all follow patterns, perhaps the universe itself is organized in a coherent way.

This idea would inspire philosophers to search for deeper harmony connecting all parts of nature.

5.5 — Harmony Beneath Complexity

As philosophers and observers looked more closely at nature, they began noticing something remarkable.

The world appeared incredibly complex on the surface. Rivers changed direction, storms formed suddenly, plants grew in different shapes, animals moved unpredictably, and the sky itself seemed filled with countless stars.

Yet beneath this complexity, patterns continued to appear.

The seasons followed each other in regular cycles. The sun and moon moved in predictable paths. Living organisms grew, matured, and reproduced according to recognizable rhythms. Even motion in the physical world seemed to follow consistent relationships.

These patterns suggested that nature might be guided by a deeper harmony.

Harmony, in this context, does not mean beauty alone. It refers to the idea that different parts of the universe might be connected through underlying relationships that maintain balance and order.

Some early thinkers began exploring this possibility.

For example, the Greek philosopher Pythagoras and his followers believed that the structure of the universe might be related to mathematical harmony. They observed that musical notes produced pleasing sounds when their vibrations followed simple numerical ratios. From this observation, they wondered whether similar numerical relationships might govern the motion of celestial bodies.

This idea led to the concept that the universe itself might resemble a form of cosmic harmony, where different motions and forces interact according to hidden patterns.

Other philosophical traditions expressed similar insights in different ways.

In ancient China, thinkers described the balance between complementary forces within nature. In Indian philosophical traditions, scholars explored how cycles of creation, preservation, and transformation shaped the living world.

Although these ideas emerged from different cultures, they shared a common intuition: the universe might not simply be a collection of separate events. Instead, it might be a system in which many parts interact in organized ways.

Recognizing harmony beneath complexity encouraged thinkers to search for deeper connections.

If the motion of the heavens follows patterns, and life itself follows cycles, perhaps these different phenomena are not isolated. They might be expressions of the same underlying order operating at multiple levels of nature.

This realization strengthened the belief that the universe is not merely understandable in fragments—it might be understandable as a whole system.

Such thinking gradually encouraged philosophers to look for principles that could describe this harmony more precisely.

And as human curiosity continued exploring these patterns, a new intellectual tool began gaining importance—one capable of describing harmony and order with extraordinary precision.

That tool was mathematics.

Closing Transition

By this stage in our journey, the human mind had already traveled far from the nights of pure storytelling.

People were no longer observing the world only with fear or wonder. They had begun noticing something deeper—patterns. The seasons returned in cycles. The sun and moon moved with remarkable regularity. Living organisms followed rhythms of birth, growth, and renewal. Even motion itself seemed to follow consistent behavior.

These discoveries slowly changed the way humans viewed the universe.

Nature was no longer just a mysterious stage of events. It appeared to possess structure. Beneath the complexity of storms, rivers, stars, and living beings, there seemed to exist a hidden order connecting everything.

Philosophers began asking an important question.

If nature follows patterns, how can we understand those patterns more clearly?

Observation had already revealed many repeating cycles, but observation alone was not always precise enough. The mind needed a way to describe patterns accurately, to compare relationships, and to communicate those relationships clearly to others.

Gradually, thinkers realized that there was a powerful language capable of doing exactly that.

Numbers.

Mathematics allowed humans to express patterns with clarity and precision. The movement of celestial bodies, the proportions of musical harmony, and the relationships between shapes and distances could all be described through numerical relationships.

For the first time, the order of nature could be written in a language that did not depend on stories or symbols.

It depended on structure.

This realization would transform philosophy once again. Thought alone would soon be joined by mathematical reasoning, creating a new way of exploring the universe.

The philosopher observing patterns in nature was slowly becoming something new.

A scientist.

In the next chapter, we explore how mathematics emerged as the language capable of describing the hidden order of the universe—turning philosophical insight into a precise tool for understanding reality.

Chapter 6 — The Idea of Cause and Effect

Understanding that events follow rules rather than divine moods

Imagine a person holding a small stone in their hand.

They lift it slightly and release it. The stone falls to the ground.

The person tries again. The same thing happens. Every time the stone is released, it falls downward. It never floats in the air, never moves upward on its own, and never disappears.

At first, such an event might seem ordinary. But when someone begins paying close attention, a deeper question emerges.

Why does the stone always fall?

Now imagine another situation.

A piece of dry wood is placed into a fire. The flames spread across the wood, producing heat and light. The wood slowly changes form, turning into glowing embers and ash.

The same event happens every time fire meets dry fuel.

Or consider rain.

Dark clouds gather in the sky. The air becomes heavy and the wind changes direction. Soon after, rain begins to fall. People living close to nature would have noticed that these events often occur together.

Certain conditions appear, and certain results follow.

At first, such events were often explained through stories. Thunder might be described as the voice of powerful beings in the sky. Rain might be interpreted as a gift or punishment from unseen forces. Fire might be seen as a mysterious element with its own personality.

These explanations helped people make emotional sense of the world.

But careful observers gradually noticed something important.

The events themselves behaved consistently, regardless of the stories told about them.

The stone fell every time it was released. Fire burned when fuel was added. Rain followed particular patterns in the sky.

These repeating relationships suggested something profound.

Events might not happen because of changing moods of invisible forces. Instead, they might occur because one event causes another.

The falling of the stone might be linked to a natural force acting on it. Fire might burn because of a process that transforms material into heat and light. Rain might follow physical conditions in the atmosphere.

Once the human mind began recognizing these connections, a powerful idea emerged.

The universe might operate through cause and effect.

Instead of random events or symbolic meanings, the world might follow relationships where certain conditions produce certain outcomes. Understanding those relationships could allow humans to explain, predict, and eventually use the forces of nature.

This realization would become one of the most important foundations of human knowledge.

Because once we believe that events follow causes, we can begin searching for the rules that govern them.

6.1 — From Event to Explanation

For most of early human history, people witnessed countless events without clearly separating the event itself from its explanation.

A storm would appear suddenly in the sky. Lightning would strike, thunder would shake the ground, and rain would fall. These events were dramatic and often frightening. Naturally, people wanted to understand why they happened.

The earliest explanations usually came in the form of stories.

If thunder frightened a community, the story might say that powerful beings in the sky were expressing anger. If a river flooded its banks, it might be explained as the will of a river spirit. Such explanations helped people give meaning to events that seemed overwhelming or mysterious.

However, as observation became more careful, some thinkers began separating two important things: the event and the explanation.

The event was what actually happened—the lightning flash, the falling rain, the flowing river.

The explanation was the story or idea used to describe why the event occurred.

This distinction opened the door to deeper inquiry.

Instead of accepting the explanation automatically, philosophers began examining whether it truly matched the pattern of the event. If thunder always followed lightning, perhaps there was a connection between the two. If certain clouds often appeared before rain, perhaps those clouds were part of the process that produced rainfall.

The mind began asking new questions.

What conditions lead to this event? Does the same event occur whenever those conditions appear? If the conditions change, does the outcome also change?

These questions slowly shifted human thinking from storytelling toward investigation.

For example, a philosopher observing fire might notice that flames grow stronger when more dry wood is added. If the wood is removed, the fire weakens. This pattern suggests that fire depends on certain materials and conditions rather than on the mood of a supernatural force.

Such observations encouraged thinkers to search for consistent relationships between events.

When the same conditions repeatedly produced the same results, it became difficult to believe that the outcome was random. Something within the structure of nature itself must be guiding the process.

This realization was the beginning of a new intellectual habit.

Instead of stopping at the event, the human mind began searching for the mechanism behind it.

Every observation became an opportunity to ask a deeper question: *What causes this to happen?*

Once this habit formed, the world began revealing its hidden connections.

6.2 — The First Recognition of Natural Rules

As thinkers continued observing nature carefully, they began noticing something important.

Certain events did not occur randomly. They followed consistent patterns that repeated again and again under similar conditions.

If dry wood was placed in a flame, the fire spread through it. If a seed was planted in fertile soil and received water and sunlight, it grew into a plant. If heavy dark clouds gathered in the sky, rain often followed.

These patterns suggested that nature might operate according to rules.

The idea of natural rules did not mean laws written somewhere in the universe. Instead, it meant that events seemed to follow reliable relationships. When specific conditions appeared, certain outcomes tended to occur.

For early observers, this realization was extremely powerful.

If the same conditions produce the same result, then the future might become predictable.

Farmers could plant crops at particular times of the year because they knew that seasonal patterns would bring suitable weather. Hunters could track animals by understanding their behavior and migration patterns. Sailors could navigate by recognizing the regular movement of stars across the sky.

In each case, human survival depended on recognizing natural rules hidden within everyday events.

Philosophers began reflecting on these patterns more deeply.

They noticed that the world seemed to behave consistently even when humans were not directly involved. The sun rose and set regardless of human beliefs. Rivers flowed downhill whether people prayed to them or not. Fire burned when fuel was present and died out when the fuel disappeared.

Such observations gradually strengthened the idea that the universe might follow principles that exist independently of human interpretation.

This realization challenged earlier explanations that relied solely on the moods or intentions of invisible forces. If events consistently followed certain conditions, then their causes might lie within the structure of nature itself.

For example, if objects always fell toward the ground when released, perhaps there was a natural force pulling them downward. If the moon followed a predictable path across the sky, perhaps its motion was governed by a consistent rule.

Although early philosophers did not yet possess the tools to discover these rules precisely, they had already recognized an important truth:

Nature behaves in ways that are reliable enough to be understood.

This belief marked a turning point in human thinking. Instead of viewing the world as a collection of unpredictable events, people began seeing it as a system governed by relationships between causes and effects.

And once the mind accepts that nature follows rules, the next step becomes inevitable.

Humans begin searching for those rules.

6.3 — Breaking the Spell of Randomness

For a long time, the world appeared unpredictable.

Storms arrived suddenly. Illness struck without warning. Crops sometimes flourished and sometimes failed. Lightning could strike a tree in an instant, and earthquakes could shake the ground without any visible cause.

To people who did not yet recognize the deeper patterns of nature, these events seemed random and frightening. It was natural to believe that powerful forces—perhaps invisible beings or divine personalities—were responsible for such unpredictable happenings.

But as observation grew more careful, a different picture began to appear.

Thinkers started noticing that events which seemed random at first often followed hidden patterns.

Rainfall did not occur completely without warning. Certain clouds, changes in wind, and shifts in air pressure often appeared before storms. Illness sometimes spread among groups of people who shared similar living conditions. Crops thrived when the soil, water, and sunlight were favorable.

These observations gradually weakened the belief that events happened purely by chance or by changing divine moods.

Instead, events began to look like processes.

A process means that one condition leads to another in a connected sequence. Clouds form under certain atmospheric conditions. Those clouds produce rain. Rain nourishes the soil. Plants grow from the nutrients in the soil.

Each step follows from the one before it.

Once thinkers began recognizing such processes, the world started appearing less mysterious and more understandable.

Lightning was no longer just a sudden act of divine power—it became something that occurred under specific atmospheric conditions. Floods followed heavy rainfall and the overflow of rivers. Disease spread through contact, environment, or other natural factors.

The universe, it seemed, was not governed by random chaos.

It followed relationships.

Recognizing these relationships allowed humans to see the world with greater clarity. Instead of feeling powerless before unpredictable forces, people could begin identifying the conditions that produced certain events.

If storms follow particular patterns, they can sometimes be anticipated. If crops grow best under certain conditions, those conditions can be created intentionally.

Breaking the spell of randomness therefore changed how humans interacted with the world.

Nature was no longer simply a mystery to be feared.

It was a system to be understood.

6.4 — Cause, Consequence, and Responsibility

The idea of cause and effect did not only change how humans understood nature. It also changed how they understood their own actions.

Once people recognized that events in nature follow certain conditions, it became easier to see that human behavior might follow similar relationships.

If planting seeds leads to crops, then failing to plant them leads to hunger. If fire is handled carefully, it provides warmth and protection; if it is handled carelessly, it can destroy homes and forests. Actions produce consequences.

This realization gradually shaped human thinking about responsibility.

In earlier times, when events were often explained through divine moods or unpredictable forces, people might attribute misfortune entirely to external causes. A failed harvest might be seen as punishment from unseen powers, or an accident might be interpreted as fate.

But the growing awareness of cause and effect introduced a different perspective.

Some outcomes might result directly from human choices.

If a community ignored signs of approaching storms, damage could follow. If food supplies were not stored for winter, hunger might result. If conflicts within a group escalated without resolution, violence could emerge.

Understanding cause and consequence therefore encouraged people to think more carefully about their actions.

Philosophers began reflecting on this connection between behavior and outcome. They explored questions about ethics, responsibility, and decision-making. If actions produce predictable consequences, then individuals and communities must consider the long-term effects of their choices.

This thinking gradually influenced the development of moral philosophy.

Many traditions around the world developed ideas emphasizing that human actions carry results. Whether expressed through ethical teachings, philosophical reflection, or cultural wisdom, the underlying insight remained similar: choices shape outcomes.

The recognition of cause and effect thus extended beyond natural events into the realm of human life.

People began realizing that the same principles guiding natural processes could also apply to social and personal behavior. Careful thinking, wise decisions, and responsible actions could help create more stable and harmonious communities.

In this way, understanding cause and effect connected two important aspects of human existence.

It helped people interpret the behavior of nature, and it helped them reflect on the consequences of their own actions.

Both insights would later influence the development of science, ethics, and social systems—fields that depend on understanding how actions lead to outcomes.

6.5 — The Foundation of Scientific Thinking

Once the idea of cause and effect became clear, a powerful transformation occurred in human thinking.

If events follow causes, then the universe is not random. It behaves in ways that can potentially be understood.

This realization created the intellectual foundation for what would later become science.

For the first time, thinkers began approaching natural events with a new assumption: if something happens, there must be a reason behind it. The task of the observer is to discover that reason.

This belief encouraged a more disciplined way of looking at the world.

Instead of accepting explanations based solely on tradition or authority, people started paying closer attention to evidence. They observed what conditions existed before an event occurred and what happened afterward. If the same conditions repeatedly produced the same outcome, it suggested a causal relationship.

Gradually, the human mind began developing a method of inquiry.

First, an event was observed. Then a possible explanation was proposed. Finally, the explanation was compared with further observations to see if it remained consistent.

Although this process was still simple compared to modern scientific methods, it represented a major step forward.

For example, if a philosopher believed that certain conditions caused rain, they might observe whether rain appeared whenever those conditions occurred. If the pattern repeated consistently, the idea gained credibility. If it did not, the explanation had to be reconsidered.

In this way, thinkers began testing ideas against the behavior of the natural world.

The belief that events follow causes also encouraged people to search for general principles. Instead of explaining each event separately, philosophers began looking for rules that could apply to many situations at once.

If motion follows certain principles, perhaps those principles apply to falling stones, flowing rivers, and moving celestial bodies alike. If plants grow under specific conditions, perhaps those conditions reveal something about the processes governing life.

This search for universal explanations gradually shaped the way knowledge developed.

Philosophy had introduced the idea that the universe could be understood through reasoning. The recognition of cause and effect strengthened that idea by suggesting that natural events are connected through discoverable relationships.

Once thinkers accepted that nature follows rules, the next challenge became learning how to identify those rules more precisely.

To achieve that, observation alone would not be enough. Measurement, experimentation, and mathematical description would soon become necessary tools.

The philosopher asking *why* was slowly becoming something new.

A scientist.

Closing Transition

With the recognition of cause and effect, the human mind had crossed another important threshold.

The universe was no longer seen as a place where events occurred according to mysterious or unpredictable moods. Instead, it began to appear as a system in which events follow relationships—where certain conditions produce certain outcomes.

This realization transformed curiosity into something more disciplined.

If natural events follow causes, then they can be studied. If they can be studied, they can be understood. And if they can be understood, humans may eventually learn how to predict them.

The idea of cause and effect therefore became one of the most powerful insights in the history of human thought.

It encouraged people to observe the world carefully, compare events, and search for patterns that reveal how nature operates. Philosophers who once asked questions about the nature of reality were now beginning to explore how events connect to one another in structured ways.

Yet an important challenge still remained.

Observation could reveal patterns, and reasoning could suggest explanations, but these tools were sometimes too vague to describe nature with precision. If thinkers wanted to understand the rules governing the universe, they needed a language capable of expressing patterns clearly and consistently.

Gradually, philosophers realized that such a language already existed.

Numbers and mathematical relationships offered a way to describe order with remarkable accuracy. Patterns in the movement of celestial bodies, the proportions found in music and geometry, and the relationships between shapes and distances could all be expressed through mathematics.

With this realization, philosophy began moving toward a new stage.

Thought would soon be joined by measurement. Reasoning would soon be joined by mathematics.

The universe was beginning to reveal not only that events follow causes, but also that these causes might be expressed through precise relationships.

And once mathematics entered the search for understanding, the next great transition would begin.

Philosophy was slowly preparing the path for science.

Reflective Interlude — The Turning of the Mind

By this point in our journey, something subtle yet powerful has already taken place in the human mind.

At the beginning, humans stood outside the cave looking at the world with curiosity and fear. Fire, storms, stars, birth, and death appeared mysterious and overwhelming. Stories helped people live with that mystery. Myths gave meaning to events that could not yet be understood.

But curiosity never stays still.

Over time, humans began noticing patterns. The seasons returned in cycles. The sun and moon moved with reliable rhythm. Life followed processes of growth and renewal. Motion appeared everywhere—in rivers, winds, flames, and the sky.

These observations slowly changed how the world was perceived.

Philosophy emerged as the first disciplined attempt to think about these patterns. Thinkers began asking questions instead of only repeating stories. They wondered whether the universe followed hidden principles that the human mind could discover.

Then came an even deeper realization.

Events are connected. One condition leads to another. Fire burns because certain materials interact. Rain follows particular atmospheric conditions. Actions lead to consequences.

The idea of **cause and effect** quietly reshaped human understanding.

The universe began to look less like a collection of unpredictable events and more like a system of relationships.

Yet philosophy still had limits.

Observation revealed patterns, reasoning suggested explanations, and discussion refined ideas—but something was still missing. Human thinking needed a way to describe patterns with precision and verify explanations more reliably.

A new stage of understanding was approaching.

Curiosity would now be joined by measurement. Reasoning would be joined by experimentation. Patterns would be described through mathematics.

This shift would give birth to a new way of exploring reality.

What began as stories became questions. Questions became philosophy. And philosophy was now preparing to become **science**.

In the next part of our journey, we will explore how humans learned to study nature systematically—transforming curiosity into one of the most powerful tools ever created by the human mind.

PART III — SCIENCE

"Nature reveals its patterns not to belief, but to observation."

Chapter 7 — Observation Changes Everything

From speculation to observation and measurement

Imagine a thinker standing beside a quiet courtyard, watching a simple event.

A stone is dropped from the hand and falls to the ground.

The thinker drops another stone. It falls again in exactly the same way.

Then a third stone. The same result.

At first, it would be easy to explain this event through a quick idea or belief. Many explanations might appear possible. Perhaps the stone falls because it prefers the earth. Perhaps the ground calls it downward. Perhaps some invisible force pulls it toward the center of the world.

But a careful observer pauses before choosing any explanation.

Instead of immediately answering *why*, the observer begins by watching more closely.

How fast does the stone fall? Does a heavier stone fall differently from a lighter one? What happens if the stone is dropped from a greater height? Does the same behavior occur every time?

This shift may seem small, but it represents one of the most important turning points in human thinking.

For centuries, philosophers had tried to understand nature mainly through reasoning and discussion. Thought experiments and logical arguments helped generate many ideas about how the universe might work. But ideas alone could not always determine which explanation was correct.

Observation introduced a new discipline.

Before explaining nature, one must first see it clearly.

Watching events carefully often reveals details that imagination alone cannot provide. Patterns appear, unexpected behaviors emerge, and assumptions sometimes prove incorrect.

For example, careful observation eventually revealed that heavier objects do not necessarily fall faster than lighter ones, contrary to what many earlier thinkers believed. Such discoveries were only possible because someone chose to observe patiently rather than rely solely on reasoning.

Observation therefore became the bridge between philosophy and science.

It grounded human curiosity in reality. Instead of asking only what might be true, thinkers began asking what actually happens when nature is examined closely.

This approach changed everything.

Once the human mind learned to observe carefully, the universe began revealing its behavior more clearly than ever before.

7.1 — The Limits of Pure Thought

For centuries, philosophers relied primarily on reasoning to understand the world. Through careful thinking and debate, they developed many ideas about the nature of reality, motion, matter, and the cosmos.

This approach achieved remarkable progress.

Philosophers learned to ask deep questions, examine contradictions, and construct explanations that seemed logically consistent. They debated the structure of the universe, the nature of change, and the causes behind natural events. These discussions sharpened the human mind and expanded intellectual curiosity.

Yet reasoning alone had an important limitation.

Ideas can be logically appealing and still be incorrect.

When thinkers rely only on thought, their explanations often reflect what seems reasonable rather than what actually occurs in nature. Human intuition can sometimes be misleading, especially when dealing with complex physical processes.

A well-known example illustrates this problem.

For many centuries, the philosopher Aristotle suggested that heavier objects fall faster than lighter ones. At first glance, this idea appears sensible. A heavier stone feels stronger in the hand than a small pebble, so it seems natural to assume that it would fall more quickly.

Without careful observation, such reasoning can easily become accepted truth.

However, when later observers tested this idea more carefully, they discovered something surprising. When dropped from the same height under similar conditions, objects of different weights often fall at nearly the same rate.

The earlier explanation, although logical, did not match reality.

This example revealed an important lesson.

Human reasoning is powerful, but it must remain connected to observation. Without checking ideas against the behavior of nature, thought can drift into speculation or reinforce incorrect assumptions.

Philosophers gradually began recognizing this limitation.

Reasoning could suggest possibilities, but observation was necessary to determine which possibility was correct. Ideas about the universe needed to be tested against the actual behavior of the world.

This realization marked a critical shift in human thinking.

Thought would remain essential for developing explanations, but it could no longer stand alone. It needed the guidance of careful observation to stay aligned with reality.

Once thinkers accepted this principle, the path toward scientific investigation became clearer.

Ideas would now be examined not only through logic, but through the evidence provided by nature itself.

7.2 — Learning to Look Carefully

Once thinkers began recognizing the limitations of pure reasoning, a new habit slowly developed: learning to observe nature carefully and patiently.

Observation may sound simple, but true observation requires discipline. It means looking at events without rushing to explain them. It means watching the same phenomenon repeatedly and noticing details that might otherwise be ignored.

Early observers began applying this approach to many natural processes.

They watched how the shadow of an object changed throughout the day as the sun moved across the sky. They noticed how the moon gradually changed shape night after night. They observed how plants responded to sunlight, water, and seasons.

Each repeated observation revealed something new.

Instead of relying on a single event, thinkers began comparing many occurrences of the same phenomenon. If a pattern appeared consistently, it suggested that the behavior might reflect a genuine property of nature rather than coincidence.

This approach required patience.

Nature does not always reveal its patterns immediately. Some cycles unfold slowly over days, months, or even years. Observers had to pay attention across long periods of time to recognize these repeating relationships.

In many cases, careful observation also corrected earlier assumptions.

For example, people might believe that certain stars remain fixed in the sky. But by observing the heavens night after night, they began noticing subtle movements in the positions of planets. These "wandering stars" behaved differently from the rest, revealing that the heavens were more dynamic than previously believed.

Similarly, watching the motion of objects helped clarify how movement actually occurs. Careful observation revealed that some assumptions about motion did not match real behavior.

Through such experiences, observers learned an important lesson.

Nature must be seen before it is explained.

Observation therefore became a central skill for anyone trying to understand the world. It grounded human thinking in reality and allowed patterns to emerge naturally from the behavior of the universe itself.

Once thinkers developed this discipline, a new question appeared.

Watching events carefully revealed patterns—but could those patterns be described more precisely?

This question led to the next great step in the evolution of knowledge: measurement.

7.3 — Measurement Enters the Picture

Careful observation revealed patterns in nature, but observation alone had an important limitation.

Two people watching the same event might describe it differently.

One observer might say that a stone falls "quickly," while another might say it falls "very quickly." Someone watching the sun move across the sky might say it moves "slowly." Such descriptions depend on personal perception, and perception is not always precise.

To understand nature more clearly, observers needed a way to describe events objectively.

This need led to one of the most important developments in human knowledge: measurement.

Measurement allows us to describe natural events using quantities rather than impressions. Instead of saying that something moves "quickly," we can measure how far it travels and how long it takes. Instead of saying that an object is "heavy," we can determine its weight using a consistent scale.

Once measurement entered the study of nature, understanding began to sharpen dramatically.

Consider motion again. When observers began measuring time and distance carefully, they could calculate how fast an object was moving. They could compare the motion of different objects and determine whether their speeds changed over time.

Such measurements revealed patterns that would have been impossible to detect through observation alone.

For example, the motion of falling objects could be analyzed more precisely by measuring how their speed increased during descent. The movement of celestial bodies could be tracked across the sky with remarkable accuracy when their positions were recorded carefully over time.

Measurement transformed vague impressions into quantitative knowledge.

To achieve this, early thinkers developed basic units for describing the world. Time could be measured using repeating natural cycles, such as the movement of the sun or the oscillation of a pendulum. Length could be compared using standardized rods or markings. Weight could be determined using balanced scales.

These tools allowed observers to record data that could be shared and verified by others.

If one observer measured the time required for an event and another repeated the same measurement, both could compare results. Knowledge was no longer dependent solely on individual interpretation—it became something that could be tested and confirmed.

Measurement therefore changed the relationship between humans and nature.

Instead of simply watching events, people could now analyze them with precision. Patterns that once appeared mysterious began revealing mathematical relationships.

This shift marked the beginning of a powerful alliance between observation and numbers.

Nature was not only observable—it was measurable.

7.4 — Instruments Extend Human Senses

As observation and measurement became more important, humans encountered another limitation.

Our senses are powerful, but they are not perfect.

The human eye can see distant mountains and bright stars, yet it cannot observe tiny organisms invisible to the naked eye. Our sense of time can estimate seconds or minutes, but it cannot measure extremely short intervals accurately. Even distance and weight are difficult to judge precisely without assistance.

To overcome these limitations, humans began creating instruments.

Instruments are tools that extend the capabilities of our senses. They allow us to observe and measure aspects of nature that would otherwise remain hidden.

One of the earliest and simplest examples is the measuring rod or ruler. By marking fixed intervals along a straight object, people could compare lengths accurately. This allowed builders, engineers, and observers to describe physical dimensions in a consistent way.

Similarly, scales and balances allowed people to measure weight. Instead of guessing whether one object was heavier than another, they could compare objects directly and determine their relative mass.

Time measurement also improved through instruments.

Early civilizations used sundials to track the movement of shadows created by the sun. Later, devices such as water clocks and mechanical clocks allowed people to measure time more reliably. With better timekeeping, observers could study motion and repeating cycles with far greater precision.

As knowledge advanced, more sophisticated instruments appeared.

The telescope allowed observers to see distant celestial bodies with unprecedented clarity. Suddenly, the moon revealed mountains and craters, and the planets showed phases and motion that had never been seen before.

The microscope opened an entirely different world.

Through magnification, observers discovered that living organisms invisible to the naked eye existed everywhere—in water, soil, and even within the human body. The discovery of microscopic life transformed our understanding of biology and medicine.

Each new instrument expanded the horizon of human knowledge.

Instead of relying solely on natural perception, scientists could now examine nature with enhanced precision and reach. Hidden structures became visible, subtle motions could be measured, and distant phenomena could be studied from Earth.

Instruments therefore played a crucial role in transforming observation into systematic investigation.

They allowed humans to explore the universe far beyond the limits of their own senses, revealing layers of reality that had previously remained unseen.

With better observation and precise measurement now available, a new possibility emerged.

If nature could be observed and measured, perhaps it could also be tested deliberately.

7.5 — From Watching to Testing

Observation allowed humans to see patterns in nature. Measurement allowed those patterns to be described with precision. But eventually, thinkers realized that one more step was necessary.

Sometimes simply watching nature was not enough.

Many natural events occur rarely or under complex conditions. Waiting for them to happen again could take a long time. Even when they did occur, it was difficult to control all the factors influencing the event.

To understand causes more clearly, thinkers began trying something new.

Instead of only observing nature as it unfolds naturally, they started creating situations where specific conditions could be controlled. This approach gave birth to what we now call experimentation.

In an experiment, the observer deliberately changes certain conditions and observes the outcome.

For example, if someone wants to understand how objects fall, they might drop objects of different weights from the same height and compare their motion. If the goal is to understand how plants grow, seeds might be planted under different conditions of sunlight, water, and soil.

By controlling conditions carefully, observers could determine which factors actually influenced the outcome.

Experiments therefore helped separate cause from coincidence.

If the same result appears every time a specific condition is created, it suggests a genuine causal relationship. If changing the condition alters the result, the observer learns something about the mechanism behind the phenomenon.

This method introduced a powerful discipline into the study of nature.

Ideas were no longer judged only by how convincing they sounded. They had to match the results observed in controlled situations. If an idea failed to match the evidence, it had to be revised or replaced.

Over time, experimentation became one of the central tools of science.

Scientists designed experiments to test hypotheses, measure outcomes, and confirm whether their explanations accurately described the behavior of nature. Through repeated testing and verification, knowledge became increasingly reliable.

Experiments also revealed something important about the universe.

Natural processes often behave consistently even when recreated in different places and times. If the same conditions produce the same results repeatedly, it suggests that nature operates according to stable principles.

This insight strengthened the belief that the universe follows discoverable laws.

Observation had revealed patterns. Measurement had quantified them. Experimentation now allowed thinkers to test their ideas against reality.

The transformation from passive watching to active testing marked the true beginning of modern scientific investigation.

Closing Transition

With observation, measurement, and experimentation now working together, human understanding had entered a new stage.

The universe was no longer approached only through speculation or philosophical debate. Nature was being studied directly—carefully observed, measured with precision, and tested through controlled experiments.

This transformation changed how knowledge itself was built.

Instead of relying on authority, tradition, or persuasive arguments, explanations now had to survive a more demanding test: they had to match the behavior of the natural world. If an idea could not explain what was observed or measured, it had to be reconsidered.

Nature had become the final judge.

Yet even with these powerful tools, another challenge remained.

Observation produced large amounts of information. Measurement generated numbers describing motion, distance, weight, and time. Experiments revealed relationships between different conditions and outcomes.

But how could these relationships be expressed clearly and consistently?

How could the patterns discovered through observation and experimentation be described in a way that allowed others to understand them, test them, and apply them?

The answer gradually emerged through a language that had already begun shaping human thought.

Mathematics.

Numbers, ratios, and equations provided a way to describe patterns with extraordinary precision. The motion of falling objects, the paths of celestial bodies, and the behavior of waves and forces could all be expressed through mathematical relationships.

With mathematics, the patterns hidden within nature could be written in a universal language.

This development marked another turning point in the journey from curiosity to knowledge.

Observation had revealed how nature behaves. Experimentation had allowed humans to test their ideas about it. Now mathematics would allow them to describe the rules governing those behaviors.

In the next chapter, we will explore how mathematics became the language of science—turning observation into equations and transforming our understanding of the universe.

Chapter 8 — Experiments and Evidence

How testing ideas created the scientific method

Imagine a thinker sitting beside a small table, observing a simple setup.

On the table lies a pendulum—a small weight attached to a string. The thinker gently pulls the weight to one side and releases it. The pendulum swings back and forth in a smooth, repeating motion.

At first glance, the movement seems ordinary. But curiosity pushes the observer to ask a question.

What determines how fast the pendulum swings?

An idea forms in the mind. Perhaps the weight of the object controls the speed of the swing. A heavier object might swing faster than a lighter one.

In earlier times, such an idea might have been accepted simply because it sounded reasonable. But now the thinker approaches the question differently.

Instead of assuming the answer, the observer decides to test the idea.

Two pendulums are prepared—one with a heavy weight and one with a lighter weight. Both are attached to strings of the same length. When released at the same time, the thinker watches their motion carefully.

To the observer's surprise, both pendulums swing with nearly the same rhythm.

The original idea turns out to be incomplete.

But this result is not a failure. It is a discovery.

By testing the idea against reality, the thinker has learned something new about how the world behaves.

This moment illustrates one of the most important developments in human knowledge: the recognition that ideas must face evidence.

An explanation might appear convincing in theory, but only careful testing can reveal whether it truly reflects the behavior of nature. Experiments allow thinkers to compare their ideas with real events, uncover mistakes, and refine their understanding.

This approach created a new discipline in human thinking.

Instead of defending ideas, scientists began examining them. Instead of relying on belief or authority, they relied on evidence produced through observation and experimentation.

Gradually, this method evolved into a systematic process for discovering truth.

An idea would be proposed. A test would be designed. Evidence would be gathered. The explanation would be accepted, modified, or rejected based on the results.

This process became the foundation of what we now call the scientific method—a powerful way of exploring the universe by allowing nature itself to answer our questions.

8.1 — From Belief to Verification

For most of human history, knowledge was closely tied to belief and authority.

Ideas about the world were often accepted because they came from respected teachers, religious traditions, or long-standing cultural explanations. If a philosopher or scholar proposed an explanation for a natural event, that idea might remain accepted for centuries simply because people trusted the authority behind it.

In many cases, these explanations were thoughtful and sophisticated. Early philosophers observed nature carefully and developed impressive systems of reasoning. Yet without systematic testing, it was difficult to determine whether an explanation truly matched reality.

Gradually, thinkers began recognizing a critical problem.

An idea may sound convincing, but that does not necessarily make it correct.

For example, someone might claim that a particular substance cures illness or that certain celestial motions control events on Earth. Such claims might appear persuasive if supported by tradition or respected figures. However, without testing the claim under controlled conditions, there is no reliable way to know whether it actually works.

This realization introduced an important shift in thinking.

Instead of asking *Who said this?*, thinkers began asking *Does it work when tested?*

Verification became more important than authority.

If an explanation predicted a certain outcome, observers could design a test to see whether that outcome occurred in reality. If the result matched the prediction, the explanation gained credibility. If the result differed, the explanation needed to be revised or abandoned.

This approach gradually changed the way knowledge developed.

Ideas were no longer protected simply because they were traditional or widely believed. They had to survive examination through observation and experiment. Even the most respected theories could be questioned if new evidence suggested a different explanation.

The transition from belief to verification therefore strengthened the reliability of knowledge.

When ideas are tested against evidence, errors become visible and corrections become possible. Understanding improves not by defending existing beliefs, but by allowing evidence to guide the search for truth.

This principle became one of the core foundations of scientific thinking.

Knowledge would no longer grow through authority alone.

It would grow through verification—the careful testing of ideas against the behavior of the natural world.

8.2 — Designing a Fair Test

Once thinkers accepted that ideas must be tested, another challenge appeared.

How should a test be designed so that the result actually reveals the truth?

A careless experiment can easily produce misleading conclusions. If too many factors change at the same time, it becomes impossible to know which one caused the result.

Early investigators gradually realized that a useful experiment must be fair and controlled.

This means that most conditions remain the same while only one factor is changed. By observing how the outcome changes when a single condition is altered, the observer can identify which factor truly influences the result.

Consider a simple example involving plant growth.

If someone wants to know whether sunlight affects how a plant grows, they might place two identical plants in similar soil and give them the same amount of water. The only difference would be that one plant receives sunlight while the other remains in shade.

If the plant receiving sunlight grows more strongly, the observer can reasonably conclude that sunlight plays an important role.

But if several conditions change at once—different soil, different water levels, and different sunlight—then the result becomes unclear. It would be difficult to determine which factor actually influenced the plant's growth.

The same principle applies in many areas of science.

When studying motion, an investigator might drop objects from the same height to compare how they fall. When studying chemical reactions, scientists carefully control temperature, pressure, and the quantities of substances involved.

Each experiment is designed to isolate the relevant factor being tested.

This approach introduced a new discipline into human inquiry. Observers had to think carefully not only about their ideas but also about how they tested them. The design of the experiment became just as important as the idea being examined.

Over time, scientists refined this practice further. They developed methods to reduce bias, improve accuracy, and ensure that results truly reflected the behavior of nature rather than errors in the experimental setup.

Through controlled testing, experiments began revealing relationships that had previously remained hidden.

This careful design of tests allowed thinkers to move beyond speculation and toward reliable understanding.

Once experiments were designed fairly, another important question naturally followed.

Even if a test produces a certain result once, how can we be sure the result is truly reliable?

8.3 — Repetition and Reliability

A single experiment can produce an interesting result, but one result alone is rarely enough to establish a reliable conclusion.

Nature is complex. Small variations in conditions, unnoticed errors, or simple coincidence can sometimes produce misleading outcomes. If an experiment is performed only once, it becomes difficult to determine whether the result reflects a genuine pattern or merely a temporary accident.

For this reason, scientists gradually adopted an essential principle: experiments must be repeated.

Repetition allows observers to see whether the same outcome appears again and again under similar conditions. If the result remains consistent across multiple trials, confidence in the explanation increases. If the outcome changes unpredictably, the original idea may need further investigation.

Consider a simple example.

Suppose an observer performs an experiment to determine how a certain chemical reaction behaves. The first test produces a particular color change. But that result alone does not guarantee that the explanation is correct.

The scientist repeats the experiment several times, carefully maintaining the same conditions. If the same color change appears every time, it suggests that the reaction follows a consistent process.

If the results differ from one trial to another, the observer must ask new questions.

Were the conditions truly identical each time? Was there an unnoticed factor influencing the outcome? Is the explanation incomplete?

Repetition therefore helps reveal the stability of natural processes.

When the same conditions consistently produce the same result, it indicates that the phenomenon is governed by a reliable relationship rather than coincidence. Such reliability allows scientists to build knowledge that others can trust.

Another important feature of repetition is that it allows independent verification.

When different observers perform the same experiment in different places and obtain similar results, confidence in the explanation becomes even stronger. Knowledge then becomes something shared across communities rather than limited to a single individual.

This practice gradually created a culture of scientific collaboration.

Scientists recorded their methods and results so that others could repeat their experiments. Over time, this collective process helped refine ideas and correct mistakes, allowing understanding to improve steadily.

Through repetition and verification, scientific knowledge gained a quality that earlier explanations often lacked.

It became reliable.

Ideas were no longer accepted simply because they sounded reasonable. They were accepted because repeated experiments demonstrated that they consistently matched the behavior of nature.

And once experiments produced reliable evidence, an even more powerful shift in thinking began to take place.

Evidence itself started carrying greater authority than tradition or reputation.

8.4 — Evidence Over Authority

As experiments became more common and results could be repeated reliably, a profound shift began to occur in the way humans treated knowledge.

For centuries, authority had played a central role in shaping understanding. The words of respected philosophers, religious scholars, or ancient texts often carried enormous influence. If a well-known thinker proposed an explanation, many people accepted it without question.

This system had advantages, especially in preserving knowledge across generations. But it also had limitations.

Even the most brilliant thinkers can make mistakes.

Once experiments and careful observation became central to scientific inquiry, a new principle gradually emerged: evidence matters more than authority.

If an experiment consistently produced results that contradicted an accepted belief, the belief itself had to be reconsidered—no matter how respected its origin might be.

One of the most famous examples of this shift occurred in the study of astronomy.

For a long time, many scholars believed that Earth stood at the center of the universe. This idea had deep roots in philosophical traditions and was widely accepted in many cultures. Yet as astronomical observations improved, some thinkers began noticing patterns in planetary motion that did not fit this model.

When new observations and calculations were examined carefully, they suggested a different possibility: the planets, including Earth, might be moving around the sun.

This idea challenged long-standing beliefs supported by authority and tradition. But the growing body of observational evidence could not easily be ignored.

Gradually, the scientific community began accepting that reliable evidence must guide understanding—even when it overturns familiar ideas.

This shift was revolutionary.

Knowledge was no longer anchored primarily in who made a claim. Instead, it depended on whether the claim could be supported by observation, measurement, and experiment.

In practical terms, this meant that every scientific explanation had to remain open to examination. If new evidence appeared, the explanation could be refined, improved, or replaced.

Rather than weakening knowledge, this process made it stronger.

By allowing evidence to challenge authority, science created a self-correcting system. Mistakes could be identified and corrected over time, leading to progressively deeper understanding of the natural world.

In this new intellectual environment, truth was not determined by reputation or tradition.

It was determined by what nature revealed through evidence.

8.5 — Building Shared Knowledge

As experiments became more systematic and evidence began guiding understanding, another important transformation occurred.

Knowledge was no longer limited to the experience of a single observer.

In earlier times, much understanding remained local or personal. A skilled craftsman might know how to work with certain materials, or a healer might possess knowledge about particular plants. Such knowledge was valuable, but it often remained confined to individuals or small communities.

Scientific investigation gradually changed this situation.

When experiments were carefully designed and repeated, their results could be recorded and shared. Other observers could read the description of the experiment, repeat it under similar conditions, and compare their results.

If the same outcome appeared again and again in different places, confidence in the explanation increased.

This process allowed knowledge to grow collectively.

Instead of relying on isolated discoveries, scientists began building upon the work of others. Observations made by one researcher could inspire new experiments by another. Ideas could be refined, corrected, or expanded through collaboration across different regions and generations.

Over time, systems were developed to support this exchange of knowledge.

Scientific writings began describing not only conclusions but also the methods used to obtain them. This transparency allowed others to verify results independently. If an experiment could not be repeated successfully, the explanation behind it would need further examination.

This practice created a new intellectual culture.

Knowledge became something that could be tested by anyone willing to follow the same method. It no longer depended solely on personal authority or isolated insight. Instead, it developed through a network of observers contributing their findings to a growing body of evidence.

Such collective effort dramatically accelerated human understanding.

Discoveries in astronomy, physics, chemistry, biology, and many other fields began accumulating over centuries. Each new generation of researchers inherited the results of earlier work and used them as a foundation for further exploration.

In this way, scientific knowledge became a shared human project.

The universe was no longer being studied by isolated individuals alone. Entire communities of thinkers were now working together to uncover the principles governing nature.

Through experiments, evidence, and shared verification, humanity had created a powerful system for discovering truth.

Closing Transition

With experimentation, evidence, and shared verification firmly established, science had gained something extraordinary—a reliable method for discovering truth.

Ideas were no longer accepted simply because they sounded convincing or came from respected authorities. Instead, explanations had to pass through a rigorous process: observation, careful testing, repeated verification, and open scrutiny by others.

Through this method, knowledge became increasingly dependable.

When the same experiment produced the same result across different places and different observers, confidence in the underlying explanation grew stronger. Over time, these reliable explanations began forming the foundations of scientific understanding.

Yet as scientists gathered more evidence through experiments, they faced another challenge.

Nature revealed patterns, but those patterns were often complex. The motion of celestial bodies, the behavior of falling objects, the interaction of forces, and the growth of waves all involved relationships that were difficult to describe using ordinary language alone.

To communicate these relationships clearly, scientists needed a language capable of expressing patterns with precision.

That language was mathematics.

Mathematics allowed scientists to describe how quantities relate to one another. The speed of a moving object could be expressed as a relationship between distance and time. The motion of planets could be represented through mathematical equations describing their paths around the sun.

Through mathematics, patterns discovered through experiments could be written in a precise and universal form.

This development transformed science once again.

Observation revealed what happens in nature. Experimentation tested ideas about those events. Mathematics now provided a way to express the laws governing those events.

With these tools combined, science began uncovering the deep structure of the universe—from the motion of planets to the behavior of matter itself.

In the next chapter, we will explore how mathematics became the language through which scientists describe the hidden order of nature, turning experimental discoveries into powerful laws that shape our understanding of the world.

Chapter 9 — The Language of Mathematics

Why mathematics became the most precise tool for describing nature

Imagine a thinker watching a simple event.

A pendulum swings gently from side to side. Back and forth, back and forth, following a smooth and steady rhythm. The motion appears calm and predictable, almost like the ticking of a clock.

At first, the observer describes what is happening in ordinary language.

The pendulum swings regularly. It moves faster at the bottom of its path. It slows down as it rises.

These descriptions capture the general behavior, but they remain somewhat vague. Words can explain the motion in a broad sense, yet they cannot express the exact pattern of the movement.

Curiosity pushes the observer to look more closely.

How long does one complete swing take? Does the weight of the pendulum affect its motion? What happens if the length of the string changes?

To answer such questions, the thinker begins measuring.

The time of each swing is recorded. The length of the string is measured. Different weights are tested. Gradually, numbers start appearing in the observations.

Something surprising begins to emerge.

When the length of the pendulum changes, the time of its swing also changes in a consistent way. The relationship between these quantities can be expressed through a precise mathematical pattern.

What once appeared as simple motion now reveals an underlying structure.

Numbers and relationships begin describing the behavior more accurately than words alone. With mathematical expressions, the observer can predict how the pendulum will behave even before testing it.

This realization has appeared repeatedly throughout the history of science.

The motion of falling objects, the path of planets around the sun, the vibrations producing musical sound, and the behavior of waves in water all follow patterns that can be described mathematically.

In each case, mathematics acts like a powerful lens.

It reveals the hidden order within natural phenomena and allows scientists to express that order with clarity and precision.

Gradually, thinkers came to understand something remarkable.

Mathematics is not merely a convenient tool invented by humans. It is a language capable of describing the structure of nature itself.

Through mathematics, the patterns discovered through observation and experiment can be written in a form that is precise, universal, and predictive.

The universe, it seems, speaks a language of relationships.

And mathematics allows the human mind to read it.

9.1 — When Words Are Not Enough

Human language is remarkably powerful. It allows us to describe experiences, share ideas, and communicate complex thoughts with others. Through words we can explain what we observe in nature—the rising of the sun, the motion of a river, the growth of a plant, or the fall of a stone.

However, when scientists began studying natural phenomena more closely, they encountered a limitation.

Words can describe what happens, but they often struggle to capture exact relationships.

Consider the motion of a falling object. One might say that the object falls "quickly" or that its speed "increases as it falls." While these descriptions provide a general sense of the event, they remain imprecise. They do not tell us exactly how fast the object moves at each moment or how its speed changes over time.

Similarly, describing the movement of planets using only words would quickly become difficult. The paths of celestial bodies involve precise relationships between distance, time, and position. Without a more exact language, these relationships remain vague.

This is where mathematics becomes essential.

Mathematics allows us to express natural phenomena through quantitative relationships. Instead of saying that an object falls quickly, we can measure how far it travels and how long the motion takes. Instead of describing a planet's motion loosely, we can calculate its position at a particular time using mathematical formulas.

Numbers transform descriptions into precise statements.

For example, if an object travels ten meters in two seconds, we can calculate its speed as five meters per second. This expression communicates far more clearly than simply saying that the object moves quickly.

The same principle applies across many areas of science.

The growth of populations, the spread of waves, the interaction of forces, and the motion of celestial bodies all involve relationships that can be described mathematically. These relationships often involve changes over time, proportions between quantities, or patterns that repeat under specific conditions.

Words alone cannot capture such details reliably.

Mathematics provides a way to express them with clarity.

This does not mean that ordinary language loses its importance. Words help us interpret ideas, communicate insights, and explain scientific concepts to others. But when it comes to describing the exact structure of natural processes, mathematics offers a level of precision that language alone cannot achieve.

As scientists explored the natural world more deeply, they discovered that many phenomena—motion, sound, light, heat, and even the behavior of matter—could be understood more clearly through numerical relationships.

Mathematics therefore became the most precise tool for describing nature.

9.2 — Discovering Numerical Patterns

As measurement became more common, scientists and thinkers began noticing something fascinating.

Many natural phenomena seemed to follow numerical relationships.

These relationships were not random. They appeared repeatedly in different areas of nature, suggesting that numbers might reflect something fundamental about the structure of the world.

One of the earliest examples appeared in music.

Ancient observers discovered that pleasant musical sounds were produced when vibrating strings followed simple numerical ratios. If the length of a string was shortened to half its original size, it produced a note one octave higher. Other harmonious sounds also emerged when strings followed simple proportional relationships.

This discovery revealed that sound and harmony could be described through numbers.

A similar pattern appeared in geometry.

Shapes such as triangles, circles, and squares displayed consistent relationships between their sides and angles. These geometric relationships could be expressed through mathematical principles that remained true regardless of the size of the shape.

Geometry therefore provided a way to understand spatial structure in nature.

Astronomy offered even more striking examples.

Careful observation of planetary motion revealed that celestial bodies followed regular paths across the sky. When astronomers began recording the positions of planets over time, they noticed patterns that could be expressed mathematically.

The distance of planets from the sun, their orbital periods, and their speeds along their paths all showed relationships that could be described through numerical laws.

Even the motion of objects on Earth displayed similar regularities.

The path of a falling object, the swing of a pendulum, and the behavior of waves on water all followed patterns that could be measured and expressed through mathematical relationships.

These discoveries led to an important realization.

Numbers were not merely tools for counting objects or measuring distances. They were revealing hidden patterns within nature itself.

The universe appeared to operate through relationships that mathematics could describe with remarkable accuracy.

Once scientists began recognizing these numerical patterns, mathematics became far more than a convenient tool. It became a powerful way to uncover the underlying structure of physical reality.

Through equations and mathematical models, scientists could identify relationships that were not immediately visible to the eye. Patterns hidden within complex phenomena began emerging through the careful analysis of numbers.

This ability transformed mathematics into one of the central languages of science.

With numbers and equations, scientists could not only describe what they observed but also begin predicting how natural systems would behave under different conditions.

And this predictive power would soon lead to some of the most profound discoveries in human history.

9.3 — Mathematics Reveals Hidden Order

As scientists began applying mathematics to natural phenomena, they discovered something extraordinary.

Mathematics did not merely describe what was already visible. It often revealed patterns that could not be seen directly through observation alone.

Many processes in nature appear complicated at first glance. The motion of planets across the sky, the path of a falling object, or the spreading of waves on water can seem irregular or difficult to understand simply by watching them.

However, when these events are measured and expressed mathematically, an underlying structure often emerges.

Consider the motion of objects.

At first, motion may appear unpredictable. A stone thrown into the air follows a curved path, slowing as it rises and speeding up as it falls. Without careful analysis, this behavior may seem difficult to describe.

But when scientists began measuring time, distance, and speed, they discovered that motion follows precise mathematical relationships. These relationships can be expressed through equations that describe how an object's position changes with time.

Similarly, the movement of planets across the sky once puzzled astronomers.

For centuries, observers carefully recorded the positions of planets night after night. When these observations were analyzed mathematically, scientists discovered that planetary motion followed specific geometric patterns.

These patterns revealed that planets travel along predictable paths governed by measurable relationships between distance, time, and gravitational influence.

Mathematics also revealed order in many other areas of nature.

The vibration of strings and air columns producing musical tones follows mathematical relationships between frequency and wavelength. Waves moving through water or air obey equations describing how energy spreads through space. The growth of populations and the flow of heat through materials can also be expressed mathematically.

In each case, mathematical analysis exposes relationships that might remain hidden without numerical representation.

This ability gives mathematics a unique power in science.

Observation allows us to see what happens. Measurement allows us to quantify it. But mathematics allows us to understand the pattern behind the phenomenon.

Through mathematical relationships, scientists can detect regularities that may not be immediately obvious. Once these relationships are discovered, they can be expressed as general principles or laws that apply to many situations.

In this way, mathematics acts like a lens that reveals the deeper structure of the natural world.

What once appeared as scattered events becomes part of an organized system governed by consistent relationships.

This realization transformed the way scientists approached nature. Instead of viewing phenomena as isolated occurrences, they began searching for mathematical patterns that unify many different observations.

And once these patterns were written in mathematical form, they became powerful tools for predicting how nature will behave under new conditions.

9.4 — Equations as Compact Descriptions of Nature

As mathematical thinking entered science, scientists discovered that complex natural processes could often be expressed through equations.

An equation is more than just a collection of numbers and symbols. It is a concise way of describing how different quantities in nature are related to one another.

Instead of explaining a phenomenon through long descriptions, an equation can capture the essential relationship in a compact form.

Consider a simple example involving motion.

If an object moves at a constant speed, the distance it travels depends on how fast it moves and how long it moves. This relationship can be written mathematically as:

distance = speed × time

With this simple equation, we can immediately calculate how far an object will travel under many different conditions. If we know any two of the quantities, we can determine the third.

This ability makes equations extremely powerful.

A well-constructed equation does not merely describe one specific event. It represents a general relationship that applies to many situations. Whether a car travels along a road or a planet moves through space, the same mathematical relationship between distance, speed, and time can help describe the motion.

As scientific understanding advanced, equations began capturing more complex patterns in nature.

For example, equations describe how objects accelerate under the influence of forces, how heat flows through materials, and how waves travel through water or air. In astronomy, equations allow scientists to predict the position of planets and comets years or even centuries into the future.

What makes equations remarkable is their predictive power.

Once a mathematical relationship is established, scientists can use it to calculate outcomes that have not yet been observed. If the equation accurately represents nature, the predictions will match real events when they occur.

This predictive ability has played a crucial role in scientific progress. Many discoveries have emerged when mathematical equations suggested phenomena that were later confirmed through observation or experiment.

Equations also provide a universal language.

Unlike ordinary language, which varies across cultures and regions, mathematical expressions can be understood by scientists around the world. An equation written in one country can be interpreted and tested by researchers elsewhere with the same meaning.

Through equations, the patterns discovered in nature become precise, compact, and universal.

In this way, mathematics allows scientists not only to describe the world but also to uncover the laws governing its behavior.

9.5 — The Power and Limits of Mathematical Models

As mathematics became central to science, it demonstrated extraordinary power.

Through equations and mathematical relationships, scientists could describe motion, energy, forces, waves, and many other natural processes with remarkable precision. Mathematical models allowed researchers to predict the behavior of planets, design bridges and machines, and understand complex interactions within physical systems.

However, while mathematics is an incredibly powerful tool, scientists gradually realized something important.

Mathematics does not replace reality—it represents it.

A mathematical model is a structured description of a natural process. It captures the relationships between certain quantities, allowing scientists to analyze and predict behavior. But every model is based on assumptions and simplifications.

For example, when studying motion, a scientist might assume that air resistance is negligible or that surfaces are perfectly smooth. These assumptions help simplify the mathematical description, making the problem easier to analyze. Yet in real situations, additional factors may influence the outcome.

This means that mathematical models must always remain connected to observation and experiment.

If an equation predicts a certain behavior but experiments consistently show something different, the model must be refined. Sometimes new variables need to be included, or a completely new framework may be required to describe the phenomenon accurately.

Scientific progress often occurs through this process.

A mathematical model successfully explains a wide range of observations, but eventually new evidence appears that the model cannot fully explain. Scientists then modify the model or develop new theories that better match the behavior of nature.

This pattern has appeared repeatedly in the history of science.

Early models of planetary motion were refined as more precise observations became available. Classical mechanics described many aspects of motion successfully, yet later discoveries in the microscopic and cosmic realms required new mathematical frameworks such as quantum mechanics and relativity.

These developments do not diminish the value of earlier models. Instead, they show that scientific understanding evolves as knowledge deepens.

Mathematics remains one of the most powerful tools for describing nature, but it must always be guided by evidence from the real world.

The strength of science lies in this balance.

Observation reveals phenomena. Experiment tests ideas about those phenomena. Mathematics describes the relationships governing them.

When these elements work together, they allow humans to explore the universe with extraordinary clarity.

Through mathematical models grounded in evidence, science continues refining our understanding of how nature operates.

Closing Transition

With mathematics now working alongside observation, measurement, and experimentation, science had finally acquired a complete toolkit for exploring nature.

Observation revealed what happens in the natural world. Experiments allowed scientists to test ideas about those events. Mathematics provided a precise language for describing the patterns discovered through these investigations.

Together, these elements formed the foundation of modern scientific understanding.

Through this combination, scientists could do something extraordinary: discover the laws governing natural phenomena. Instead of studying each event in isolation, they could identify general principles that applied across many different situations.

For example, the same mathematical laws that describe the motion of a falling object on Earth can also describe the motion of planets around the sun. The equations governing waves in water help explain the behavior of sound and light. Relationships discovered in small laboratory experiments can sometimes describe processes occurring on a cosmic scale.

This unifying power revealed that the universe might operate according to a small number of fundamental principles expressed through mathematical relationships.

Once these principles were discovered, their implications became enormous.

Scientists could predict the motion of celestial bodies, understand the behavior of matter, and design machines that harness natural forces. Knowledge was no longer limited to explaining events after they occurred—it could now anticipate what would happen before it happened.

This predictive power transformed science into a practical force.

Understanding natural laws allowed humans to build technologies that reshaped civilization. From navigation and engineering to medicine and communication, scientific knowledge began influencing nearly every aspect of human life.

In this way, the journey from curiosity to mathematics completed a remarkable transition.

What began as stories about nature gradually became philosophical questions. Those questions evolved into careful observation and experimentation. Mathematics then allowed those discoveries to be expressed as precise laws.

The next step in this journey would reveal how these laws began transforming the world around us.

Science was no longer only a way of understanding nature.

It was becoming a powerful tool for engineering reality itself.

Reflective Interlude — When the Universe Became Readable

We have now reached a remarkable turning point in the journey of human understanding.

At the beginning of our story, humans looked at the world with wonder and fear. Thunder, fire, stars, birth, and death were powerful experiences, but they were difficult to explain. Stories and myths helped early humans live with this mystery. They provided meaning and comfort in a world that often seemed unpredictable.

Then curiosity began asking deeper questions.

Philosophy emerged when humans started wondering not only *what happens*, but *why it happens*. Thinkers searched for patterns in nature, recognizing that the world might follow hidden principles rather than unpredictable moods of unseen forces.

The next transformation came when observation became disciplined. Instead of relying only on reasoning, people began watching nature carefully, measuring events, and comparing results. Experiments allowed ideas to be tested directly against reality.

Finally, mathematics entered the picture.

With numbers and equations, the patterns of nature could be expressed with extraordinary precision. Motion, waves, planetary paths, and many other phenomena began revealing consistent relationships that could be written in mathematical form.

For the first time in human history, the universe appeared readable.

Nature was no longer just a collection of mysterious events. It behaved like a vast system governed by relationships that could be observed, tested, and expressed mathematically.

This realization gave birth to modern science.

Yet everything we have explored so far—the development of observation, experimentation, and mathematical description—was only the preparation.

The real adventure was about to begin.

Once scientists had the tools to study nature systematically, they began uncovering some of the most profound truths ever discovered by the human mind. These discoveries revealed the hidden structure of the cosmos, the laws governing motion and energy, the nature of matter, and the processes that shape life itself.

Each discovery expanded humanity's understanding of the universe and transformed how we live.

In the next part of our journey, we will explore the great scientific breakthroughs that emerged once humans learned how to read the language of nature.

PART IV — THE GREAT DISCOVERIES

"Each discovery is a moment when the universe allows us to see one more piece of its design."

Chapter 10 — Motion and Gravity

From falling apples to planetary motion

Imagine standing beneath a tree on a quiet afternoon.

A ripe apple detaches from a branch and falls to the ground. The event lasts only a moment. It is so ordinary that most people would hardly notice it. Apples fall from trees every day. Stones drop when released. Rain falls from clouds.

For most of human history, such events were simply accepted as part of nature.

But a curious mind sees something different in this simple moment.

Why does the apple fall downward instead of drifting sideways or floating upward? What determines the path it follows? Is the fall of the apple connected to other motions we observe in the world?

Once such questions appear, the ordinary event begins to reveal a deeper mystery.

Look up at the night sky.

The moon moves steadily across the heavens. Planets wander slowly among the stars. These celestial bodies remain suspended in space, following predictable paths that repeat over time.

Why do they not fall to Earth the way the apple does?

Or perhaps they are falling—but in some different way.

For centuries, thinkers struggled with these questions. The motion of objects on Earth and the motion of celestial bodies in the sky were often treated as separate mysteries. One belonged to the everyday world of falling objects and moving bodies. The other belonged to the distant realm of the heavens.

But at some point, a bold idea emerged.

What if the same principle governs both?

What if the force that pulls the apple toward the ground also guides the motion of the moon and the planets?

This simple question opened the door to one of the greatest insights in the history of science.

The fall of an apple was no longer just a small event beneath a tree.

It became a clue to understanding the motion of the entire universe.

10.1 — The Puzzle of Motion

Motion is one of the most familiar features of the natural world.

Objects fall toward the ground. Rivers flow downhill. Wind moves across the land. Animals walk, run, and fly. In the sky, the sun rises and sets each day while the moon and stars follow their paths across the heavens.

Yet for centuries, understanding the rules governing motion proved surprisingly difficult.

Early philosophers observed motion carefully, but many of their explanations were based mainly on reasoning rather than systematic experimentation. One influential view, proposed by Aristotle, suggested that objects move according to their natural tendencies.

According to this idea, heavy objects naturally move downward toward the Earth, while lighter substances such as air and fire move upward. In this framework, motion required a continuous cause. If the cause stopped, the motion would stop as well.

This explanation appeared reasonable at first.

When a cart is pushed, it moves. When the push stops, the cart eventually slows down and comes to rest. Observations like these seemed to support the idea that motion requires a constant force.

However, this explanation could not fully account for all observations.

For example, what keeps the moon moving across the sky? What maintains the motion of planets as they travel along their paths year after year?

If motion requires a constant push, what provides that push in the heavens?

These questions puzzled thinkers for generations.

The difficulty arose partly because everyday experiences on Earth are influenced by factors such as friction and air resistance. These forces slow moving objects and make it appear as though motion naturally fades away unless something continues to push.

Without recognizing these hidden influences, it was difficult to uncover the true principles governing motion.

Another challenge involved measurement.

Before the development of accurate clocks and systematic experiments, observers could only describe motion in general terms—objects moved "fast" or "slow," but the exact relationship between speed, distance, and time remained unclear.

As scientific thinking advanced, researchers began examining motion more carefully.

They measured how objects fall, how far they travel, and how their speed changes over time. These observations gradually revealed patterns that earlier thinkers had overlooked.

The puzzle of motion was slowly being solved.

What had once appeared as a simple everyday phenomenon was about to become one of the most precisely understood aspects of the natural world.

10.2 — Galileo and the Study of Falling Objects

The puzzle of motion began to unravel when scientists started studying it through careful experiments and measurement rather than relying only on reasoning.

One of the most influential figures in this transformation was Galileo Galilei.

Galileo realized that understanding motion required observing it under controlled conditions. Instead of accepting traditional explanations, he began performing simple but insightful experiments to examine how objects actually move.

One common belief at the time was that heavier objects fall faster than lighter ones. This idea seemed intuitive. A heavy stone feels stronger in the hand than a small pebble, so it appeared reasonable to assume that the heavier object would fall more quickly.

Galileo questioned this assumption.

To investigate the matter, he studied the motion of falling objects in a systematic way. Rather than simply dropping objects from great heights—where motion occurs too quickly to measure easily—Galileo slowed the process by rolling balls down inclined planes.

This clever method allowed him to observe motion more carefully.

By measuring how far the balls traveled and how long the motion took, Galileo discovered something surprising: objects do not simply fall at a constant speed. Instead, their speed increases gradually during the fall.

In other words, falling objects accelerate.

This discovery was crucial because it revealed a precise pattern behind motion. The distance traveled by a falling object increases in a predictable way as time passes. Motion was not random—it followed mathematical relationships.

Galileo also demonstrated that, in the absence of air resistance, objects of different masses fall at nearly the same rate. This result challenged long-standing beliefs and showed the importance of testing ideas through experiment.

Beyond falling objects, Galileo studied many other aspects of motion. He examined the behavior of pendulums, the motion of projectiles, and the role of inertia—the tendency of objects to maintain their motion unless acted upon by external forces.

These insights laid the foundation for a new understanding of mechanics.

Galileo's work demonstrated that motion could be studied scientifically through observation, measurement, and mathematics. By combining these tools, he transformed motion from a philosophical puzzle into a measurable physical process.

His discoveries prepared the way for an even greater breakthrough.

A few decades later, another scientist would take Galileo's insights and connect them to the motion of the entire universe.

10.3 — Newton's Insight: The Universal Law of Gravity

The work of Galileo had revealed important truths about motion on Earth. Objects accelerate as they fall, and their motion follows precise mathematical patterns. Yet a larger mystery still remained.

The heavens seemed to obey different rules.

Planets moved across the sky in graceful paths. The moon circled the Earth month after month without falling. Comets appeared and disappeared along enormous arcs through space. These celestial motions looked orderly, but their cause was still unclear.

Then came one of the most profound insights in the history of science through the work of Isaac Newton.

Newton asked a simple but revolutionary question:

Could the same force that pulls objects toward the Earth also act across the vast distances of space?

If an apple falls toward the ground, perhaps the Earth is pulling it. But if the Earth can pull an apple, might it also pull the moon?

Newton imagined that the moon might actually be falling toward the Earth continuously, but because it is also moving sideways through space, it never reaches the ground. Instead, it keeps curving around the Earth in a circular path.

This idea connected two phenomena that had previously seemed unrelated: the fall of objects on Earth and the motion of celestial bodies in the sky.

To describe this connection, Newton formulated the universal law of gravitation.

According to this law, every object in the universe attracts every other object with a force that depends on two things:

- the mass of the objects
- the distance between them

Larger masses produce stronger gravitational attraction, while greater distance weakens the force.

With this single principle, Newton unified the physics of Earth and the physics of the heavens.

The same force that causes an apple to fall also governs the motion of the moon around the Earth and the motion of planets around the sun. Gravity acts everywhere, linking every object in the universe through a network of attraction.

Even more remarkable was Newton's use of mathematics.

By combining his laws of motion with the law of gravitation, he could calculate the paths of planets and explain the observations of astronomers with astonishing accuracy. The motion of the cosmos, which had once seemed mysterious, now followed precise mathematical relationships.

This achievement transformed science.

For the first time, a single set of laws could explain both everyday motion on Earth and the vast movements of celestial bodies. The universe appeared not as a collection of unrelated phenomena but as a coherent system governed by universal principles.

The falling apple and the orbiting moon were no longer separate mysteries.

They were expressions of the same fundamental law.

10.4 — A Universe Governed by Laws

Newton's discoveries did more than explain why objects fall or why planets move through space. They revealed something far deeper about the nature of the universe.

Before these discoveries, many people believed that the heavens and the Earth operated according to different principles. The sky was often considered a perfect and mysterious realm, while the Earth was viewed as a separate world filled with change and imperfection.

Newton's work shattered this division.

By showing that the same gravitational force governs both falling objects on Earth and the motion of celestial bodies, he demonstrated that the same physical laws apply everywhere in the universe.

This realization was revolutionary.

The universe was no longer divided into separate domains governed by different rules. Instead, it appeared as a single system operating according to consistent principles.

An apple falling from a tree, the moon orbiting the Earth, and the planets traveling around the sun were all part of the same physical framework.

This insight introduced the concept of universal laws of nature.

A law of nature is a consistent relationship describing how physical systems behave under certain conditions. Such laws do not change from one place to another or from one time to another. They apply equally on Earth and across the vast distances of space.

The idea that nature follows universal laws strengthened confidence in the scientific method.

If the same principles operate everywhere, then observations made in one place can help explain phenomena occurring far away. Experiments performed on Earth can reveal insights about the behavior of distant celestial bodies.

This unification of knowledge encouraged scientists to search for other universal principles.

If gravity governs motion across the cosmos, perhaps other forces also follow similar rules. Perhaps the behavior of light, heat, electricity, and magnetism can be described through laws that apply throughout the universe.

Newton's work therefore marked a turning point in human understanding.

Nature was no longer viewed as a collection of isolated events. It was seen as an interconnected system governed by consistent principles that could be discovered through observation, experimentation, and mathematics.

For the first time, the universe appeared ordered in a universal way.

10.5 — Predicting the Motion of the Cosmos

Once the laws of motion and gravity were understood mathematically, scientists gained an extraordinary new ability.

They could predict how objects in the universe would move.

Before this breakthrough, the movement of planets and celestial bodies had to be observed continuously in order to understand their behavior. Astronomers carefully recorded the positions of planets night after night, building tables of observations that helped them anticipate future positions.

Newton's laws changed this situation completely.

By combining his laws of motion with the law of universal gravitation, scientists could calculate how celestial bodies would move under the influence of gravitational forces. The motion of a planet around the sun could be described through mathematical equations that accounted for the attraction between the two bodies.

With these equations, astronomers could determine where a planet would appear in the sky months or even years into the future.

This predictive power was astonishing.

Comets that once seemed unpredictable could now be tracked along their long paths through the solar system. Scientists could calculate when a comet would return after decades or centuries of absence. The motion of moons around their planets could also be predicted with remarkable precision.

Even small irregularities in planetary motion could be analyzed.

In some cases, astronomers noticed tiny deviations in the predicted paths of planets. Instead of dismissing these deviations as errors, scientists used the mathematical laws of gravity to investigate further.

One famous example involved the discovery of the planet Neptune. Astronomers noticed that the motion of Uranus did not perfectly match predictions based on known gravitational influences. By analyzing these deviations mathematically, scientists concluded that another unseen planet must be exerting gravitational influence.

Calculations were performed to determine where this hidden planet should be located.

When astronomers pointed their telescopes toward the predicted region of the sky, they discovered Neptune almost exactly where the calculations had indicated.

This event demonstrated the extraordinary power of mathematical laws in science.

Scientists were no longer limited to observing what was already visible. They could use mathematical reasoning to predict the existence and behavior of objects that had not yet been directly observed.

Through these developments, gravity and motion became powerful tools for exploring the cosmos.

The universe revealed itself as a system whose behavior could be described, calculated, and predicted through the laws of nature.

Closing Transition

The discovery of the laws of motion and gravity marked one of the greatest turning points in the history of science.

For the first time, the motion of objects on Earth and the motion of celestial bodies in space were understood through the same set of principles. The fall of an apple, the orbit of the moon, and the paths of planets around the sun were no longer separate mysteries.

They were all governed by the same universal law.

This realization revealed that the universe behaves like a coherent system governed by consistent relationships. Once these relationships were expressed mathematically, scientists gained the ability not only to explain natural phenomena but also to predict them with remarkable accuracy.

The success of these laws inspired enormous confidence in the scientific method.

If gravity and motion could be understood through mathematical principles, perhaps other natural phenomena could be explained in similar ways. Scientists began searching for laws governing heat, light, electricity, magnetism, and the behavior of matter itself.

This search soon produced another major breakthrough.

During the eighteenth and nineteenth centuries, researchers began exploring mysterious forces associated with electric charges and magnetic fields. At first, electricity and magnetism appeared to be separate phenomena. Yet careful experiments gradually revealed that they were closely connected.

These discoveries would lead to a new understanding of energy and fields—ideas that would transform technology and reshape modern civilization.

In the next chapter, we will explore how scientists uncovered the relationship between electricity and magnetism, revealing another powerful force that shapes the behavior of the natural world.

Chapter 11 — Light and the Nature of Reality

Understanding vision, optics, and electromagnetic waves

Every morning when we wake up and open our eyes, the world suddenly appears around us.

Walls, trees, people, the sky, the sunlight coming through a window—everything becomes visible almost instantly. We rarely stop to think about how extraordinary this experience actually is.

What allows us to see?

At first glance, vision feels natural and effortless. But if we pause for a moment, the process becomes mysterious. The world does not glow on its own. Objects around us are often dark or silent. Yet somehow they become visible the moment light reaches them.

Imagine a beam of sunlight entering a dark room through a small opening in the wall.

In the beam, tiny particles of dust float slowly through the air. They become visible only when the light touches them. Outside the beam, the same dust remains invisible.

This simple observation reveals something profound.

Light makes things visible.

But what exactly is light?

For centuries, thinkers struggled with this question. When we see an object, is something leaving our eyes and touching the object? Or is something coming from the object into our eyes?

Ancient philosophers proposed different explanations. Some believed that the eye sends out rays that interact with objects, allowing us to perceive them. Others suggested that objects emit something that travels to the eye.

Gradually, through observation and experiment, scientists discovered that light travels from a source, reflects from objects, and enters our eyes, allowing the brain to form an image of the world.

Yet this explanation opened even deeper questions.

How does light travel across space so quickly? Why does it reflect from mirrors and bend through glass? What is its true nature?

When scientists began studying light carefully, they realized that it was far more than a simple tool for vision. Light carried information about distant stars. It revealed the structure of materials. It even hinted at deeper principles governing the behavior of energy in the universe.

What began as a simple question—*how do we see?*—soon became one of the most important investigations in the history of science.

Through the study of light, scientists would uncover profound insights about reality itself.

11.1 — The Mystery of Seeing

For most of human history, vision appeared so natural that people rarely questioned how it actually worked.

We open our eyes and immediately perceive shapes, colors, movement, and depth. The world appears before us almost instantly. Because this experience feels effortless, early thinkers found it difficult to understand the mechanism behind it.

One of the earliest puzzles involved a simple but important question:

How does the eye receive information about distant objects?

Ancient philosophers proposed different answers.

Some thinkers believed that vision occurs because the eyes send out invisible rays that touch objects. According to this idea, the eye actively reaches outward toward the world, allowing us to perceive things at a distance. When the eye's rays encounter an object, the mind becomes aware of it.

At first, this explanation seemed reasonable. After all, we can turn our gaze toward something and immediately see it, as though the act of looking itself produces vision.

However, other philosophers suggested a different possibility.

Perhaps objects themselves emit something that travels toward the eye. In this view, the eye acts as a receiver rather than a transmitter. Light from an object enters the eye, and the brain interprets this information to form an image.

To resolve this debate, careful observation became necessary.

Thinkers noticed that vision depends strongly on light sources. In darkness, even objects located directly in front of us remain invisible. When light illuminates them, they suddenly become visible again.

This observation suggested that light plays a crucial role in vision.

Another important clue came from shadows and reflections. When light strikes an object, it can reflect from the surface and travel in new directions. If these reflected rays enter the eye, the object becomes visible.

Gradually, experiments confirmed that vision occurs when light from a source travels to an object, reflects from it, and enters the eye.

Inside the eye, this light passes through a lens that focuses it onto the retina, a sensitive layer at the back of the eye. The retina converts the incoming light into electrical signals that travel to the brain, where they are interpreted as images.

This discovery marked a major step in understanding vision.

Seeing is not simply a mysterious ability of the mind—it is a physical process involving light, reflection, and the interaction of energy with the human eye.

Once scientists recognized this connection, the study of vision became closely linked with the study of light itself.

Understanding how light behaves—how it travels, reflects, and bends—became essential for explaining how we perceive the world around us.

11.2 — Light as Rays: The Birth of Optics

Once scientists understood that light enters the eye and enables vision, the next question became clear:

How does light travel?

Careful observation revealed a simple but powerful clue. When light passes through a small opening into a dark room, it forms a narrow beam. Dust particles floating in the air become visible inside this beam, showing the path along which light travels.

These observations suggested that light moves in straight lines.

This idea became the foundation of optics, the scientific study of light and its behavior.

One of the earliest and most important observations involved reflection. When light strikes a smooth surface such as a mirror, it bounces away in a predictable direction. By measuring the angles involved, scientists discovered a simple rule:

The angle at which light strikes a surface equals the angle at which it reflects away.

This principle allowed people to understand how mirrors form images. When light from an object reflects off a mirror and enters the eye, the brain interprets the path of the reflected rays as though they came from behind the mirror, creating the illusion of an image.

Another important phenomenon involved refraction.

When light passes from one material into another—for example, from air into water or glass—it changes direction. This bending occurs because light travels at different speeds in different materials.

Anyone who has looked at a straight stick partially submerged in water may have noticed that the stick appears bent at the water's surface. This visual effect occurs because light rays from the submerged portion of the stick bend as they pass from water into air.

Scientists studied this behavior carefully and discovered that the bending of light follows precise mathematical relationships.

Understanding reflection and refraction led to important technological developments.

Lenses could be designed to focus light and form clear images. By shaping pieces of glass carefully, scientists created instruments that extended human vision far beyond its natural limits.

Two of the most transformative inventions based on optical principles were the telescope and the microscope.

Telescopes allowed astronomers to observe distant planets, stars, and galaxies, revealing the vast structure of the universe. Microscopes opened an entirely new world, allowing scientists to see tiny organisms and cellular structures invisible to the naked eye.

Through the study of light as rays, optics became one of the first areas where observation, experiment, and mathematics worked together to reveal the behavior of nature.

Yet as experiments became more refined, scientists discovered that the ray model could not explain everything about light.

Some phenomena suggested that light might behave in a very different way—more like a wave moving through space.

11.3 — Waves and the Behavior of Light

The idea that light travels in straight lines explained many phenomena—reflection from mirrors, the formation of shadows, and the focusing of images by lenses. The ray model of light was extremely useful for understanding optics and building instruments such as telescopes and microscopes.

However, as experiments became more precise, scientists began encountering phenomena that the simple ray model could not fully explain.

One of the most intriguing discoveries involved interference.

When two beams of light overlap under certain conditions, they do not simply add together. Instead, they produce patterns of alternating bright and dark regions. In some places the light becomes stronger, while in other places it seems to cancel out entirely.

This behavior suggested something unusual.

If light were composed only of particles traveling in straight lines, such patterns would be difficult to explain. But if light behaved like a wave, the phenomenon made sense.

Waves have a property called interference. When two waves meet, their peaks and valleys combine. If two peaks meet, the resulting wave becomes stronger. If a peak meets a valley, they may cancel each other out.

Another phenomenon supporting the wave nature of light is diffraction.

When light passes through a very narrow opening or around the edge of an object, it spreads out instead of continuing strictly in a straight line. This spreading creates patterns of light and shadow that resemble the behavior of waves in water.

These observations led scientists to propose that light behaves as a wave traveling through space.

According to the wave theory, light consists of oscillations that move outward from a source, much like ripples spreading across the surface of a pond when a stone is dropped into water.

Experiments in the nineteenth century provided strong evidence for this idea. Carefully designed experiments produced interference and diffraction patterns that matched the predictions of wave theory with remarkable accuracy.

The wave model explained many aspects of light's behavior.

It accounted for interference patterns, diffraction effects, and the way light interacts with certain materials. It also helped scientists understand color as a property related to the wavelength of light waves.

Shorter wavelengths correspond to colors such as violet and blue, while longer wavelengths correspond to colors such as red.

For a time, the wave theory appeared to offer a complete explanation of light.

Yet an even deeper mystery remained.

If light is a wave, what exactly is waving?

Water waves move through water. Sound waves travel through air. But light can travel through empty space, even across the vast vacuum between the sun and the Earth.

This puzzle led scientists to search for a deeper understanding of the nature of light.

Their investigation would soon reveal a surprising connection between electricity, magnetism, and light itself.

11.4 — Electricity, Magnetism, and Light

By the nineteenth century, scientists had begun studying two mysterious phenomena that appeared unrelated to light: electricity and magnetism.

Electricity was observed when certain materials produced sparks, attracted small objects, or generated currents flowing through wires. Magnetism was known from natural magnets and compasses that could influence pieces of iron and align themselves with Earth's magnetic field.

At first, these effects seemed completely separate from one another.

Electricity appeared to involve moving charges, while magnetism seemed associated with certain materials and directional forces. Yet careful experiments gradually revealed that these two phenomena were closely connected.

When electric current flows through a wire, it creates a magnetic field around the wire. Similarly, changing magnetic fields can produce electric currents in nearby conductors.

This discovery revealed that electricity and magnetism are not independent forces—they are different aspects of a single phenomenon known as electromagnetism.

Scientists studying these interactions eventually developed mathematical equations describing how electric and magnetic fields behave and interact.

Among the most important contributions came from James Clerk Maxwell.

Maxwell's equations described how electric and magnetic fields change and influence one another. When these equations were analyzed, they revealed something astonishing.

Disturbances in electric and magnetic fields can travel through space as waves.

Even more surprising, the speed predicted for these waves matched the known speed of light.

This remarkable result led to a profound conclusion:

Light itself is an electromagnetic wave.

In other words, the light we see with our eyes is simply one form of energy created by oscillating electric and magnetic fields moving through space.

This insight unified several areas of physics that had previously seemed unrelated. The study of electricity, magnetism, and optics were suddenly part of a single framework describing electromagnetic phenomena.

Understanding light as an electromagnetic wave also explained how it can travel through empty space.

Unlike sound waves, which require air or another medium to propagate, electromagnetic waves consist of interacting electric and magnetic fields that sustain one another as they move forward. This allows light to travel across the vast vacuum between stars and planets.

The discovery of electromagnetism transformed science and technology.

It explained the behavior of light, paved the way for electrical engineering, and eventually led to inventions such as radio communication, radar, and wireless technology.

Yet this discovery also revealed something even more surprising.

The visible light we perceive with our eyes is only a small part of a much larger family of electromagnetic waves.

11.5 — Expanding the Invisible Spectrum

Once scientists understood that light is an electromagnetic wave, a new question naturally arose.

If light is simply one form of electromagnetic radiation, could there be other types of electromagnetic waves that we cannot see?

The answer turned out to be yes.

The light visible to human eyes occupies only a very small portion of a much broader range of electromagnetic waves known as the electromagnetic spectrum.

Visible light includes the familiar colors of the rainbow—from red through orange, yellow, green, blue, and violet. These colors correspond to different wavelengths of electromagnetic waves. Red light has a longer wavelength, while violet light has a shorter one.

However, beyond these visible wavelengths lie many other forms of electromagnetic radiation.

On the side of longer wavelengths, scientists discovered infrared radiation. Although invisible to the human eye, infrared waves can be detected as heat. Many modern technologies, such as thermal cameras, rely on detecting infrared radiation emitted by objects.

Beyond infrared waves lie radio waves, which have even longer wavelengths. Radio waves are used extensively in communication systems, including radio broadcasting, television transmission, and wireless communication.

On the other side of the visible spectrum lie waves with shorter wavelengths.

Just beyond violet light lies ultraviolet radiation, which carries more energy than visible light and plays an important role in processes such as sterilization and certain chemical reactions.

Still further along the spectrum are X-rays, capable of passing through soft tissues while being absorbed by denser materials such as bone. This property makes X-rays extremely useful in medical imaging.

At the highest-energy end of the electromagnetic spectrum are gamma rays, which are produced in nuclear reactions and certain cosmic processes. These waves carry enormous energy and can penetrate deeply into matter.

The discovery of the electromagnetic spectrum revealed that reality contains many layers beyond what human senses can perceive directly.

Our eyes detect only a narrow band of electromagnetic radiation. Yet the universe continuously emits and absorbs waves across a vast range of wavelengths and energies.

With the help of scientific instruments, humans have learned to observe these invisible forms of radiation. Radio telescopes detect signals from distant galaxies. Infrared sensors reveal hidden heat patterns. X-ray detectors allow doctors to examine the internal structure of the human body.

Each new region of the spectrum opened another window into the universe.

Through these discoveries, scientists realized that the world we perceive through our senses represents only a small portion of a much richer physical reality.

The study of light therefore transformed our understanding of nature.

It showed that reality extends far beyond what we can see—and that scientific tools allow us to explore these hidden layers of the universe.

Closing Transition

The study of light changed our understanding of reality in a profound way.

At first, light appeared to be simply the medium that allows us to see the world around us. But as scientists investigated its behavior more carefully, it revealed something much deeper.

Light obeys precise laws of reflection and refraction. It behaves like a wave, capable of interference and diffraction. It is also part of a larger electromagnetic phenomenon linking electricity and magnetism. And the visible light we perceive represents only a small portion of a vast electromagnetic spectrum.

These discoveries expanded the boundaries of human perception.

Through instruments and scientific investigation, humanity learned that the universe contains many layers beyond what the eye can directly observe. Heat radiation, radio signals, X-rays, and gamma rays all carry information about physical processes occurring throughout the cosmos.

The world, it seemed, was far richer and more complex than our senses alone could reveal.

Yet as scientists explored the behavior of light and electromagnetic waves, they encountered new puzzles that existing theories could not fully explain.

In some experiments, light behaved clearly like a wave. In others, it appeared to behave more like tiny packets of energy interacting with matter in discrete ways. This unexpected behavior suggested that the nature of light might be even more subtle than previously imagined.

At the same time, scientists studying matter began uncovering surprising properties of atoms and particles. The behavior of energy and matter at very small scales did not always follow the familiar rules of classical physics.

These discoveries would soon lead to one of the most revolutionary developments in the history of science.

In the next chapter, we will explore how scientists began uncovering the quantum nature of reality, revealing a universe far stranger and more fascinating than anyone had previously imagined.

Chapter 12 — Energy, Heat, and Matter

Fire, engines, and the birth of thermodynamics

Long before humans built machines or studied science, they discovered something that would transform their lives forever.

Fire.

The first controlled flames brought warmth during cold nights, protection from predators, and the ability to cook food. Fire allowed early humans to survive in harsh environments and gradually expand across the planet. It also made possible the shaping of tools, the hardening of clay into pottery, and later the melting of metals.

For thousands of years, people used fire without truly understanding what it was.

They could see the flames, feel the heat, and observe how fire transformed materials—wood turning into glowing embers, metals softening in a furnace, food changing its texture and flavor when cooked.

These transformations suggested that fire carried some powerful force.

Early thinkers proposed different explanations for this phenomenon. Some believed that heat was a special substance released during burning. Others imagined that fire represented one of the fundamental elements of nature.

Yet these explanations did not fully explain what heat actually is or how it works.

A new level of curiosity emerged during the early stages of industrial development.

When engineers began building machines powered by steam, they encountered a fascinating puzzle. Heat from burning fuel could be used to boil water, producing steam that pushed pistons and turned wheels. In other words, heat could be converted into mechanical motion.

This raised important questions.

What exactly is heat? How can heat produce motion? Is there a general principle connecting heat, motion, and other forms of energy?

These questions led scientists to study heat in a systematic way.

Careful experiments began revealing that heat was not a mysterious substance but rather a form of energy associated with the motion of particles within matter. Understanding this connection allowed scientists to develop the field now known as thermodynamics.

Thermodynamics explores how energy moves, transforms, and flows through physical systems. It explains how engines work, why heat flows from hot objects to colder ones, and why certain processes can occur while others cannot.

The study of heat and energy therefore became one of the central pillars of modern science.

From the simple flame of a campfire to the powerful engines of industrial machines, the investigation of heat revealed deep principles governing how matter and energy interact throughout the universe.

12.1 — Fire: Humanity's First Energy Tool

Among all the discoveries made by early humans, few were as transformative as fire.

Before the control of fire, human survival depended almost entirely on daylight and natural shelter. Nights were dark and cold, and protection against predators was limited. The ability to create and maintain fire changed this situation dramatically.

Fire provided warmth, allowing humans to inhabit colder regions of the Earth. It produced light, extending activity beyond sunset. Perhaps most importantly, it enabled cooking.

Cooking food made it easier to digest and extract nutrients, which may have played an important role in human development. Over time, the use of fire also helped communities preserve food, prepare medicines, and improve living conditions.

Beyond survival, fire opened the door to early technology.

By heating materials, humans discovered that substances could be transformed in remarkable ways. Clay hardened into pottery when placed in a kiln. Certain rocks melted to produce metals such as copper, bronze, and later iron. These discoveries marked the beginnings of metallurgy and craftsmanship.

Each of these transformations involved energy released through fire.

Yet for thousands of years, people used fire without fully understanding what was happening inside the flame. Fire appeared as a mysterious process that consumed materials while producing heat and light.

Some early thinkers believed that fire was one of the fundamental elements of nature. Others imagined that burning released a hidden substance within materials.

These ideas reflected attempts to explain an observation that everyone could see but no one fully understood.

Despite the lack of scientific explanation, the practical power of fire was undeniable. From heating homes to forging tools and weapons, fire became one of the most important energy sources in human civilization.

Over time, as societies became more complex, humans began using fire not only for survival but also for mechanical work.

The burning of fuel could heat water, producing steam that pushed pistons and drove machines. This discovery would eventually lead to the development of steam engines and the rise of industrial technology.

But before engineers could fully harness the power of heat, scientists needed to understand what heat actually was and how it behaved.

This challenge led to one of the most important debates in the history of science—the true nature of heat itself.

12.2 — The Mystery of Heat

As scientists began studying fire and heat more carefully, they realized that understanding heat was not as simple as it first appeared.

Everyone could feel heat. When a metal rod is placed in a flame, it becomes hot. When two objects touch, heat may flow from the warmer one to the cooler one. Yet the nature of heat itself remained uncertain.

One influential explanation proposed during the eighteenth century was known as the caloric theory.

According to this idea, heat was considered a special invisible fluid called *caloric*. This fluid was believed to flow from hotter objects to colder ones. When a substance contained more caloric, it was warmer. When caloric flowed out, the object cooled down.

The caloric theory seemed to explain many everyday observations.

For example, when a hot object touches a cold one, heat appears to flow from the hot object to the cold one until both reach the same temperature. This behavior resembled the movement of a fluid from a region of higher concentration to a region of lower concentration.

Because this model explained several phenomena reasonably well, it remained widely accepted for many years.

However, some observations began raising doubts.

One important clue came from experiments involving friction. When two surfaces rub against each other, they generate heat. If heat were truly a fixed fluid stored inside materials, it would be difficult to explain how friction could seemingly produce unlimited amounts of heat.

A famous example involved the drilling of cannon barrels. During the drilling process, large quantities of heat were generated continuously as the metal was bored. The amount of heat produced did not appear to diminish even after long periods of drilling.

This observation suggested that heat might not be a stored fluid after all.

Instead, some scientists proposed a different explanation: heat might be related to motion.

According to this idea, heat arises from the microscopic motion of particles within matter. When particles move rapidly and collide with one another, the material becomes warmer. When their motion slows, the temperature decreases.

This view gradually gained support through further experiments.

Scientists discovered that mechanical work—such as friction or compression—could be converted into heat. Conversely, heat could sometimes be transformed into mechanical motion, as seen in steam engines.

These discoveries revealed that heat was not a mysterious substance.

It was a form of energy associated with the motion of particles.

Understanding this connection between motion and heat laid the groundwork for a new field of science—one that would explore how energy flows, transforms, and governs physical processes throughout nature.

12.3 — Engines and the Industrial Revolution

While scientists were debating the nature of heat, engineers were confronting a very practical challenge.

How can heat be used to produce mechanical work?

This question became extremely important during the Industrial Revolution. Factories, mines, and transportation systems required machines capable of performing heavy tasks more efficiently than human or animal labor.

One of the earliest and most powerful solutions came in the form of the steam engine.

In a steam engine, fuel such as coal is burned to heat water. The heated water turns into steam, and the expanding steam pushes against a piston inside a cylinder. This motion can then be used to drive wheels, pumps, or other mechanical devices.

Steam engines soon powered factories, locomotives, and ships, transforming transportation and industry.

However, early steam engines were inefficient.

Large amounts of fuel were required to produce relatively small amounts of useful work. Engineers wanted to improve the performance of these machines, but doing so required a deeper understanding of how heat actually behaves.

This practical problem attracted the attention of scientists.

One important figure in this investigation was Sadi Carnot. Carnot examined how heat flows through an engine and how this flow can produce mechanical motion.

He realized that an engine operates by transferring heat from a hot source to a cold environment. During this process, part of the energy from heat can be converted into useful work.

Carnot also discovered an important limitation.

No engine can convert all of its heat into work. Some energy must always be lost as heat flowing to the cooler surroundings. This insight revealed that natural processes place fundamental limits on the efficiency of machines.

These investigations transformed the study of heat from a collection of observations into a systematic science.

Scientists began recognizing that heat, motion, and work are deeply connected. Mechanical motion can generate heat through friction, and heat can produce motion in engines. Both are different forms of energy interacting with matter.

The need to understand engines therefore helped scientists uncover general principles governing energy transformation.

These principles would soon become known as the laws of thermodynamics—fundamental rules describing how energy behaves in physical systems.

12.4 — The Laws of Thermodynamics

As scientists studied heat engines, mechanical work, and the behavior of energy in different systems, they began discovering general principles that applied to all energy transformations.

These principles became known as the laws of thermodynamics.

Thermodynamics is the branch of science that studies how energy moves, changes form, and interacts with matter. Through careful experiments and theoretical analysis, scientists realized that energy follows certain universal rules that cannot be violated.

The first law of thermodynamics expresses one of the most fundamental ideas in physics: energy is conserved.

Energy cannot be created or destroyed. It can only change from one form to another.

For example, when fuel burns in a steam engine, the chemical energy stored in the fuel is converted into heat. Some of that heat energy becomes mechanical motion as steam pushes a piston. Other portions may become sound, vibration, or heat released into the surrounding environment.

Although the forms of energy change, the total amount remains the same.

This law revealed that the universe operates like a vast system in which energy constantly transforms but never disappears.

The second law of thermodynamics introduced another important insight.

While energy is conserved, not all energy can be converted into useful work. In every energy transformation, some energy spreads into forms that are less useful for performing mechanical tasks.

For example, when a machine operates, friction generates heat that disperses into the surrounding environment. Once this heat spreads out, it becomes difficult to recover and use again.

The second law therefore describes the natural tendency of energy to move from concentrated forms to more dispersed forms.

Scientists often describe this process using the concept of entropy, a measure of the disorder or spreading of energy within a system. As natural processes occur, entropy tends to increase.

This principle explains why certain processes happen spontaneously while others do not. Heat naturally flows from hot objects to colder ones, but it does not spontaneously flow in the opposite direction without external work being applied.

Together, these laws revealed that energy transformations follow strict constraints.

Machines cannot operate with perfect efficiency. Some energy always disperses as heat, and natural processes tend to move toward states where energy becomes more evenly distributed.

Understanding these laws had enormous practical importance.

They guided engineers in designing engines, refrigerators, power plants, and many other technologies that rely on controlled energy transformations.

More broadly, thermodynamics showed that energy governs transformation throughout nature—from the burning of fuel to the functioning of living organisms and the evolution of physical systems.

12.5 — Energy as a Universal Concept

As thermodynamics developed, scientists gradually realized that energy is not limited to heat or engines.

Energy is a universal concept that connects many different processes occurring throughout nature.

Mechanical motion involves energy when objects move or change position. Heat represents energy associated with the motion of particles inside matter. Chemical reactions release or absorb energy when atoms rearrange their bonds. Electrical currents carry energy through moving charges. Even light and electromagnetic waves transport energy through space.

These diverse phenomena all share a common feature: they involve the transfer or transformation of energy.

This realization unified many areas of science that had previously been studied separately.

For example, when a fuel burns in an engine, chemical energy stored in the fuel is converted into heat. That heat produces steam, which creates mechanical motion in the engine's piston. The piston may then drive a generator that produces electricity.

In this single process, energy changes form several times—chemical energy becomes heat, heat becomes mechanical motion, and motion becomes electrical energy.

The same principle applies in living systems.

Plants capture energy from sunlight through photosynthesis and convert it into chemical energy stored in organic molecules. Animals consume these molecules as food, transforming chemical energy into motion, body heat, and biological activity.

In both machines and living organisms, energy continuously flows and transforms while obeying the laws of thermodynamics.

Recognizing energy as a universal concept allowed scientists to describe physical processes within a unified framework.

Instead of studying heat, motion, electricity, and chemical reactions as completely separate phenomena, researchers began viewing them as interconnected expressions of energy interacting with matter.

This perspective proved extremely powerful.

It allowed scientists to analyze complex systems ranging from engines and power plants to ecosystems and planetary climates. In each case, understanding how energy flows through the system reveals how the system behaves.

Thermodynamics therefore did more than explain heat engines.

It revealed that energy lies at the heart of all physical transformations.

Every process in the universe—from the burning of a flame to the shining of stars—involves the movement and transformation of energy.

Closing Transition

The study of heat, engines, and energy revealed something profound about the universe.

Energy is not confined to a single phenomenon such as fire or motion. It appears in many forms—heat, mechanical work, electricity, chemical reactions, and radiation—and these forms can transform into one another while obeying universal laws.

Thermodynamics showed that these transformations follow strict principles.

Energy is conserved, constantly changing form but never disappearing. At the same time, natural processes tend to disperse energy, making some transformations irreversible. These insights explained why engines have limits, why heat flows from hot objects to cold ones, and why certain processes occur naturally while others require external work.

Through these discoveries, scientists realized that energy governs transformation throughout nature.

From the burning of fuel to the operation of industrial machines, from the metabolism of living organisms to the dynamics of stars, energy flows through every system in the universe.

Yet as scientists studied energy and matter more closely, they began encountering new mysteries.

Experiments in chemistry and physics revealed that matter itself is not continuous or uniform. Instead, substances appear to be composed of extremely small building blocks—particles far too tiny to see directly.

These particles were called atoms.

Understanding the structure of atoms would become one of the most important challenges in science. As researchers investigated the behavior of matter at very small scales, they discovered phenomena that classical physics could not fully explain.

These discoveries would lead to a new revolution in scientific thought.

In the next chapter, we will explore how scientists uncovered the atomic structure of matter, revealing a hidden world of particles and interactions that shape the physical universe.

Chapter 13 — Electricity and the Invisible World

How unseen forces transformed civilization

Imagine entering a dark room and pressing a small switch on the wall.

Within a fraction of a second, the room fills with light. Machines begin operating, screens illuminate, and invisible signals start flowing through wires and circuits. The modern world awakens instantly at the touch of a button.

Electricity has become so deeply woven into everyday life that we rarely pause to think about it.

Cities glow throughout the night because of electrical lighting. Trains and elevators move because of electric motors. Messages travel across continents through electrical networks that carry signals at extraordinary speeds. Computers, communication systems, and countless devices rely on controlled flows of electric energy.

Yet only a few centuries ago, electricity was not a practical tool at all.

It appeared as a strange and mysterious phenomenon.

Early observers noticed unusual effects when certain materials were rubbed together. Tiny sparks could jump between objects. Lightweight pieces of straw or dust would suddenly move as though attracted by an invisible force. Lightning in the sky demonstrated that powerful electrical forces existed in nature, but their origin and behavior remained unknown.

These observations fascinated early scientists.

What was this invisible force that could attract objects, produce sparks, and sometimes release enormous energy in the form of lightning?

Unlike fire or mechanical motion, electricity could not be easily seen or touched. Its presence had to be inferred through its effects—movement of objects, flashes of light, or shocks felt by the human body.

The mystery of electricity gradually drew scientists into a deeper investigation of invisible forces.

Through careful experiments, they began uncovering the behavior of electric charges, currents, and fields. What started as a curious natural phenomenon eventually became one of the most powerful tools in human civilization.

The exploration of electricity did more than produce new technologies.

It revealed that the universe contains powerful forces operating beyond the limits of human senses—forces that shape both the structure of matter and the functioning of modern society.

13.1 — The First Observations of Electricity

Long before electricity powered cities and machines, it appeared only as a strange and curious phenomenon.

One of the earliest observations was recorded in ancient Greece. When a piece of amber—a fossilized tree resin—was rubbed with cloth or fur, it gained the ability to attract small, lightweight objects such as bits of straw, dust, or feathers.

This effect seemed mysterious.

Nothing visible connected the amber and the small objects, yet they moved toward it as if pulled by an invisible force. The Greek word for amber was *elektron*, and from this word the modern term electricity eventually emerged.

For centuries, such observations remained little more than curiosities.

People could reproduce the effect by rubbing materials together, but they had no clear explanation for why it happened. The phenomenon appeared weak and unpredictable, useful mainly for demonstration rather than practical application.

During the seventeenth century, scientists began studying these effects more carefully.

One of the early investigators was William Gilbert. Gilbert performed systematic experiments with various materials and discovered that many substances—not just amber—could produce similar attractive forces when rubbed.

He called these materials electrics, distinguishing them from substances that did not show such behavior.

Gilbert's work represented an important shift. Instead of treating electrical effects as mysterious curiosities, he began analyzing them as natural phenomena that could be studied through observation and experiment.

Over time, scientists realized that rubbing materials together could produce electric charge.

Objects could become positively or negatively charged, and these charges could interact in predictable ways. Similar charges repel each other, while opposite charges attract.

Another remarkable discovery followed.

When enough electric charge accumulates, it can suddenly discharge in the form of a spark. This spark resembles a tiny version of lightning. Observing this similarity led scientists to suspect that lightning itself might be a large-scale electrical phenomenon occurring in the atmosphere.

These early investigations transformed electricity from a simple curiosity into a subject worthy of serious scientific study.

What began with a piece of rubbed amber gradually evolved into a deeper exploration of invisible forces that operate throughout nature.

13.2 — Discovering Electric Charges and Currents

As experiments with electricity became more systematic, scientists began uncovering patterns in the behavior of electric phenomena.

One of the most important discoveries involved the idea of electric charge.

Through repeated observations, researchers noticed that electrically affected objects could behave in two distinct ways. Some objects repelled each other, while others attracted one another. To explain this behavior, scientists proposed that electric charge exists in two types.

These types were eventually named positive and negative.

Objects carrying the same type of charge repel each other, while objects with opposite charges attract. This simple rule helped explain many electrical effects that had previously seemed mysterious.

Another important question soon followed:

How can electric charge move?

Early experiments showed that when two charged objects were connected by certain materials, the charge could flow from one object to another. This flow of charge became known as an electric current.

The discovery of electric currents opened an entirely new area of investigation.

Scientists began exploring how charges move through different materials. Some substances, such as metals, allow electric charges to move easily. These materials are called conductors. Other materials resist the movement of charge and are known as insulators.

Understanding the difference between conductors and insulators made it possible to control the flow of electricity.

Another breakthrough came with the invention of the electric battery, which provided a steady source of electric current. Instead of relying on temporary charges created by friction, scientists could now produce continuous electrical flows for experiments.

One of the earliest batteries was developed by Alessandro Volta.

Volta constructed a device consisting of alternating layers of metal and chemical solutions. This arrangement produced a continuous electrical current, allowing scientists to perform controlled experiments on electricity for the first time.

With reliable sources of current available, researchers began studying how electricity interacts with matter.

They discovered that electric currents can produce heat, cause chemical reactions, and generate magnetic effects. These observations revealed that electricity was not merely a curious force—it was a powerful phenomenon capable of interacting with many aspects of the physical world.

As investigations continued, scientists uncovered an even more surprising relationship.

Electric currents could produce magnetic fields, and magnetism could in turn influence electric currents. This discovery would soon unite two previously separate areas of science.

13.3 — Magnetism Joins the Story

For centuries, magnetism had been known as a separate and fascinating phenomenon.

Natural magnets, found in certain minerals such as lodestone, could attract pieces of iron. Compasses built from magnetized needles helped sailors navigate across oceans by aligning with Earth's magnetic field.

Despite these practical uses, magnetism seemed unrelated to electricity.

Electricity appeared as sparks and charges produced by friction or chemical reactions. Magnetism appeared as a property of certain materials and the Earth itself. For a long time, scientists studied these two phenomena separately.

This situation changed dramatically in the early nineteenth century.

One of the turning points came through an experiment performed by Hans Christian Ørsted.

Ørsted noticed something surprising while demonstrating electrical currents. When an electric current flowed through a wire, a nearby compass needle moved slightly from its usual north–south direction.

The compass needle was responding to the electric current.

This simple observation revealed something extraordinary: electric currents create magnetic effects.

Electricity and magnetism were not separate forces after all. They were deeply connected.

Scientists quickly began investigating this relationship in greater detail.

Further experiments showed that the magnetic influence around a current-carrying wire forms circular patterns surrounding the wire. The moving electric charges generate a magnetic field in the space around them.

Another breakthrough soon followed.

Researchers discovered that the relationship also works in the opposite direction. Changing magnetic fields can produce electric currents in nearby conductors. This phenomenon became known as electromagnetic induction.

One of the pioneers of this discovery was Michael Faraday.

Faraday demonstrated that moving a magnet near a coil of wire could generate an electric current in the wire. This meant that motion within a magnetic field could produce electricity.

These discoveries unified electricity and magnetism into a single framework known as electromagnetism.

The consequences were enormous.

If electric currents can produce magnetic fields and magnetic motion can generate electric currents, then electricity can be used to create motion—and motion can be used to produce electricity.

This understanding laid the foundation for two of the most important technologies of modern civilization: electric motors and electric generators.

Through these developments, the invisible forces of electricity and magnetism began transforming from scientific curiosities into powerful tools capable of reshaping the modern world.

13.4 — Electricity Becomes Technology

Once scientists understood the relationship between electricity and magnetism, a new possibility emerged.

Electricity could be controlled and harnessed to perform useful work.

This realization marked the beginning of one of the most transformative technological revolutions in human history.

One of the first practical developments was the electric generator. Using the principle of electromagnetic induction discovered by Michael Faraday, engineers learned how to produce electricity by moving conductors within magnetic fields.

In a generator, mechanical motion—such as the rotation of a turbine—causes coils of wire to move through a magnetic field. This motion produces electric currents that can be transmitted through wires to power devices.

The reverse process also proved possible.

If electricity flows through a conductor within a magnetic field, it can produce motion. This principle led to the development of the electric motor, a machine that converts electrical energy into mechanical movement.

Electric motors soon appeared in factories, transportation systems, and household appliances. Machines that once depended on steam engines or manual labor could now operate more efficiently using electric power.

Another revolutionary development was the telegraph.

Before electrical communication, messages had to travel physically—carried by messengers, ships, or horses. Communication across long distances could take days or even weeks.

With the telegraph, electrical signals transmitted through wires could send coded messages across vast distances almost instantly. This technology dramatically accelerated communication and played an important role in commerce, journalism, and government coordination.

Lighting technology also underwent a transformation.

Early cities relied on candles, oil lamps, and gas lighting, which were often dim and hazardous. The development of the electric light bulb allowed homes, streets, and workplaces to be illuminated more safely and efficiently.

One of the pioneers in practical electrical lighting systems was Thomas Edison.

Edison and other engineers helped design electrical power networks capable of generating electricity in central stations and distributing it through wires to homes and businesses.

As these technologies spread, electricity began reshaping everyday life.

Factories could operate more efficiently, transportation systems became more advanced, and communication networks expanded across entire continents.

What began as a mysterious spark produced by rubbing materials together had evolved into a powerful technological force driving modern civilization.

13.5 — The Invisible Network of Civilization

As electrical technologies matured, electricity gradually became the circulatory system of modern civilization.

Unlike many earlier technologies, electricity itself is largely invisible. Wires hidden inside walls, underground cables, and transmission lines stretching across landscapes quietly carry energy from power plants to cities, homes, and industries.

Yet the impact of this invisible flow is everywhere.

Modern cities depend on vast electrical networks that operate continuously. Power stations generate electricity using sources such as coal, water turbines, nuclear reactions, wind, or sunlight. This electrical energy then travels through high-voltage transmission lines to distribution networks that deliver it to individual buildings and devices.

When electricity arrives at its destination, it can perform many different functions.

It powers lights that illuminate streets and homes. Electric motors run elevators, trains, fans, and countless machines used in industry and manufacturing. Heating and cooling systems regulate indoor environments. Medical equipment in hospitals depends on reliable electrical supply to support life-saving procedures.

Communication systems also rely heavily on electricity.

Signals traveling through electrical circuits enable telephones, broadcasting systems, computers, and global internet networks. Electrical energy allows information to move across continents in fractions of a second, connecting billions of people around the world.

Transportation has also been transformed.

Electric trains and subways move millions of passengers daily in large cities. Electric vehicles are becoming increasingly common as societies seek more efficient and sustainable forms of transportation.

Even many systems that appear mechanical or digital are ultimately dependent on electrical energy.

Computers process information using electrical signals moving through microscopic circuits. Data centers storing vast quantities of information operate entirely through electrical infrastructure. Satellites orbiting Earth rely on electrical systems powered by solar panels.

Through these networks, electricity quietly supports nearly every aspect of modern life.

Because it can be transmitted efficiently and converted into many other forms of energy—light, motion, heat, and information—electricity has become one of the most versatile tools ever developed by humanity.

What once appeared as a mysterious spark between objects has evolved into a vast global system that sustains modern civilization.

Yet the study of electricity also led scientists to ask deeper questions about the nature of matter itself.

If electricity involves moving charges, what exactly carries these charges inside materials?

Answering this question would lead scientists into an even deeper exploration of the hidden structure of atoms.

Closing Transition

The discovery of electricity did more than provide humanity with a powerful technological tool.

It opened a window into an invisible world of forces operating beneath the surface of everyday reality.

For centuries, people had understood motion, heat, and light mainly through visible phenomena. Electricity revealed that powerful interactions could exist even when they could not be directly seen. Wires carrying electric current appear ordinary, yet within them enormous numbers of tiny charges move in coordinated ways, producing light, motion, and communication across vast distances.

This realization pushed scientists to ask deeper questions.

If electricity is a flow of charges, what exactly are these charges made of? Where do they exist within matter? Why do some materials conduct electricity while others resist it?

To answer these questions, researchers had to look beyond visible matter and explore the internal structure of substances.

Gradually, experiments revealed that matter is not continuous. Instead, it is composed of extremely small building blocks known as atoms. Within these atoms exist even smaller components responsible for carrying electric charge.

Understanding these tiny structures would transform science once again.

The study of atomic structure would explain chemical reactions, electrical conduction, and many properties of materials. At the same time, it would reveal surprising behaviors of matter and energy at extremely small scales.

These discoveries would eventually lead scientists into the strange and fascinating domain of modern atomic and quantum physics.

In the next chapter, we will explore how scientists uncovered the hidden architecture of matter—revealing a microscopic universe of atoms, particles, and interactions that shape the physical world.

Chapter 14 — The Quantum Surprise

When matter and reality behaved differently than expected

By the end of the nineteenth century, many scientists believed that the major mysteries of nature had already been solved.

The laws of motion explained how objects move. Gravity described the motion of planets and falling bodies. Thermodynamics revealed how heat and energy transform. Electromagnetism unified electricity, magnetism, and light under a single framework.

The universe appeared orderly and predictable.

If the position and motion of every object were known precisely, it seemed possible that the future behavior of the entire system could be calculated. Nature looked like a vast machine governed by precise mathematical laws.

Yet beneath this confidence, certain experiments began producing results that did not fit within the existing framework.

Scientists studying heat radiation noticed that hot objects emit light in very specific patterns that classical physics could not explain. Experiments with light interacting with metals produced strange results that contradicted wave-based theories. Investigations into atomic behavior revealed patterns that seemed impossible under classical assumptions.

These anomalies were not small technical errors.

They were signs that something fundamental about nature was not yet understood.

At first, scientists tried to adjust existing theories to explain these observations. But the more experiments were performed, the clearer it became that the classical picture of reality was incomplete.

Nature at extremely small scales behaved in unexpected ways.

Energy sometimes appeared in discrete amounts rather than continuous flows. Light sometimes behaved as though it were made of tiny particles. Even particles of matter seemed capable of behaving like waves.

These discoveries challenged centuries of scientific intuition.

The familiar ideas of motion, energy, and matter had to be reconsidered.

Gradually, scientists began developing a new framework capable of explaining these strange observations. This new theory became known as quantum mechanics.

Quantum theory did not simply refine classical physics.

It revealed that the microscopic world operates according to principles that can appear deeply counterintuitive when compared with everyday experience.

The universe, it seemed, was far stranger—and far more fascinating—than anyone had previously imagined.

14.1 — The Limits of Classical Physics

By the late nineteenth century, classical physics appeared extraordinarily successful.

The laws developed by scientists such as Isaac Newton described the motion of objects with remarkable accuracy. The theory of electromagnetism developed by James Clerk Maxwell explained the behavior of light and electromagnetic waves. Thermodynamics revealed the principles governing heat, energy, and engines.

Together, these theories created a powerful framework for understanding the natural world.

Engineers used them to design machines, build bridges, develop electrical systems, and predict astronomical events with extraordinary precision. Many scientists believed that physics was nearing completion—that only small details remained to be filled in.

However, certain experiments began revealing troubling inconsistencies.

One of the first problems involved thermal radiation. When objects are heated, they emit light and other forms of electromagnetic radiation. Scientists attempted to calculate how this radiation should behave using classical physics.

The predictions did not match experimental observations.

According to classical theory, a heated object should emit enormous amounts of energy at very short wavelengths, eventually producing an infinite amount of radiation. This prediction clearly contradicted reality. Real objects emitted radiation in specific patterns that classical theory could not explain.

This puzzle became known as the ultraviolet catastrophe.

Another mystery appeared in experiments involving light interacting with matter. When light strikes certain metal surfaces, it can eject electrons from the material. Classical wave theory predicted that the energy of these electrons should depend on the intensity of the light.

Experiments showed something different.

The energy of the ejected electrons depended on the frequency of the light rather than its intensity. Increasing the brightness of low-frequency light did not produce the expected effect, contradicting classical predictions.

At the same time, scientists studying atoms encountered additional puzzles.

Atoms produced very specific patterns of light called spectral lines. Instead of emitting light across a continuous range of wavelengths, atoms emitted light only at certain precise frequencies.

Classical physics offered no clear explanation for this behavior.

These problems indicated that the established theories of physics were incomplete when applied to extremely small scales.

The successes of classical physics remained valid for many everyday situations, but new experiments were revealing that the microscopic world followed different rules.

These anomalies forced scientists to reconsider some of their deepest assumptions about matter, energy, and the structure of reality.

A new way of thinking would soon emerge—one that would reshape our understanding of the universe.

14.2 — Energy Comes in Packets

One of the first clues that nature behaves differently at very small scales came from the study of thermal radiation.

When objects are heated, they emit energy in the form of electromagnetic radiation. A glowing piece of metal, for example, first appears dull red, then orange, and eventually bright white as its temperature increases. Scientists wanted to understand exactly how this radiation depends on temperature and wavelength.

By the late nineteenth century, physicists had developed mathematical formulas based on classical physics to describe this process. However, these formulas produced a serious problem.

According to classical theory, as the wavelength of radiation becomes shorter and shorter, the amount of energy emitted by a heated object should increase without limit. In other words, a hot object should release enormous—and theoretically infinite—amounts of radiation at very short wavelengths.

This prediction clearly did not match reality.

Actual measurements showed that radiation from hot objects follows a very specific pattern. The intensity increases up to a certain point and then decreases at shorter wavelengths. The catastrophic explosion of energy predicted by classical theory simply does not occur.

This contradiction became known as the ultraviolet catastrophe.

To resolve this puzzle, the German physicist Max Planck proposed a bold new idea in 1900.

Planck suggested that energy is not emitted or absorbed continuously.

Instead, energy is exchanged in small discrete units, which he called quanta.

According to this idea, a system cannot release just any arbitrary amount of energy. It can only emit energy in specific packets determined by the frequency of radiation involved.

At first, Planck introduced this concept mainly as a mathematical trick that allowed his equations to match experimental results. Even he was uncertain whether these energy packets represented something physically real.

Yet the implications were profound.

The idea of quantized energy meant that nature does not always behave in a smooth, continuous manner. Instead, certain physical processes occur in discrete steps.

This was a radical departure from the classical view of physics, which assumed that energy could vary continuously.

Planck's idea of quantization would soon influence other scientists studying light and matter.

It suggested that the microscopic world might follow rules very different from those governing everyday experience.

14.3 — Light Behaves Like Particles

At the beginning of the twentieth century, most scientists believed that the nature of light had already been settled.

After the work of James Clerk Maxwell, light was understood as an electromagnetic wave. Experiments showing interference and diffraction strongly supported this wave model. It explained many optical phenomena with remarkable accuracy.

Yet certain experiments began revealing behavior that the wave theory could not fully explain.

One of the most puzzling examples involved the photoelectric effect.

In these experiments, scientists shined light onto a metal surface and observed that electrons were emitted from the material. According to classical wave theory, increasing the intensity of light should provide more energy to the electrons, eventually causing them to escape from the metal.

However, experiments showed something surprising.

Electrons were emitted only when the light had a frequency above a certain threshold. Increasing the brightness of low-frequency light did not produce the expected effect, no matter how intense the light became.

Even more strangely, when the frequency of the light increased, the energy of the emitted electrons increased as well.

These results could not be explained by classical wave theory.

In 1905, Albert Einstein proposed a bold explanation.

Einstein suggested that light behaves as though it consists of tiny packets of energy, later called photons.

Each photon carries a specific amount of energy determined by the frequency of the light. When a photon strikes an electron in a metal, it transfers its energy to the electron. If the energy is large enough, the electron can escape from the material.

This explanation successfully accounted for the experimental observations.

Low-frequency light consists of photons with insufficient energy to free electrons, regardless of how many photons arrive. High-frequency light carries photons with greater energy, allowing electrons to escape more easily.

Einstein's idea revealed something astonishing.

Light, which had long been described as a wave, could also behave as a collection of particles interacting with matter.

This discovery introduced the concept of wave–particle duality, suggesting that light possesses both wave-like and particle-like properties depending on how it is observed.

The implications were profound.

If light could behave both as a wave and as particles, perhaps other forms of matter might also exhibit similar dual behavior.

This possibility soon led scientists to investigate the behavior of particles themselves—opening the door to another surprising discovery.

14.4 — Matter Behaves Like Waves

After the discovery that light could behave both as a wave and as particles, scientists began asking an even more surprising question.

If light—once believed to be purely a wave—can behave like particles, could particles of matter also behave like waves?

At first, this idea seemed almost impossible.

In everyday life, matter appears solid and well defined. A stone, a grain of sand, or a metal object follows predictable paths when it moves. Classical physics describes these motions using Newton's laws, treating objects as particles with definite positions and velocities.

However, the new insights from quantum theory suggested that nature at very small scales might behave differently.

In 1924, the French physicist Louis de Broglie proposed a remarkable hypothesis.

He suggested that every particle of matter possesses wave-like properties.

According to de Broglie, particles such as electrons are not simply tiny solid objects. They are associated with waves whose wavelength depends on the particle's momentum.

This idea initially seemed abstract, but soon experiments confirmed it.

When beams of electrons were directed at certain materials, they produced interference and diffraction patterns—the same types of patterns that waves produce when they pass through narrow openings or interact with obstacles.

Such patterns cannot be explained if electrons behave purely as classical particles.

Instead, they demonstrate that electrons can act as waves spreading through space.

This discovery transformed the understanding of matter.

Electrons, atoms, and other microscopic particles do not always behave like tiny billiard balls. Under certain conditions, they behave like waves capable of interference and superposition.

These wave properties play a crucial role in the structure of atoms.

Electrons inside atoms do not simply orbit the nucleus like planets around the sun. Instead, they exist in wave-like states that determine where they are most likely to be found. These states form the basis of atomic structure and explain why atoms emit specific patterns of light known as spectral lines.

The idea that matter can behave like waves challenged many traditional assumptions about physical reality.

Particles could no longer be described solely by precise positions and trajectories. Instead, their behavior had to be described using probabilities and wave-like patterns.

This realization forced scientists to develop an entirely new mathematical framework for understanding the microscopic world.

14.5 — A New Picture of Reality

As evidence accumulated that energy and matter behave in quantized and wave-like ways, scientists realized that the classical picture of reality could no longer describe the microscopic world.

A new framework had to be developed.

During the 1920s, several physicists created the foundations of what is now known as quantum mechanics. One of the pioneers was Erwin Schrödinger, who introduced equations describing how matter waves evolve through space and time. Around the same period, Werner Heisenberg proposed another approach that described physical systems using mathematical relationships between measurable quantities.

Although these approaches looked different mathematically, they led to the same predictions.

Quantum mechanics revealed that particles such as electrons do not follow precise, deterministic paths like planets or billiard balls. Instead, their behavior is described by probabilities.

A quantum system is represented by a wave-like mathematical function that indicates where a particle is most likely to be found. When a measurement is made, the system appears in one of the possible states described by this function.

This idea introduced a profound shift in scientific thinking.

Classical physics assumed that every object has a definite position and velocity at all times. Quantum mechanics showed that at microscopic scales, these properties cannot always be known simultaneously with perfect precision.

This limitation became known as the uncertainty principle, formulated by Heisenberg. It states that the more precisely the position of a particle is known, the less precisely its momentum can be determined, and vice versa.

The quantum framework also explained several previously mysterious phenomena.

It clarified why atoms emit light at specific wavelengths, why electrons occupy particular energy levels within atoms, and how chemical bonds form between atoms to create molecules.

In other words, quantum mechanics revealed the underlying principles governing the structure of matter itself.

Although the mathematical language of quantum theory can be complex, its implications are profound.

The world at the smallest scales does not behave like the predictable mechanical universe imagined by classical physics. Instead, it follows a set of rules that involve probability, quantization, and wave-like behavior.

This new picture of reality reshaped modern science and laid the foundation for many technologies that define contemporary life—from semiconductors and lasers to modern electronics and quantum computing.

The quantum surprise revealed that nature, when examined closely enough, behaves in ways that challenge our deepest intuitions about how the universe works.

Closing Transition

The discovery of quantum behavior transformed our understanding of nature.

What began as small inconsistencies in experimental observations gradually revealed that the microscopic world operates according to principles very different from those governing everyday experience. Energy appears in discrete units, light behaves both as waves and particles, and matter itself exhibits wave-like properties.

These insights forced scientists to move beyond the classical picture of the universe.

Instead of a perfectly predictable mechanical system, the microscopic world revealed patterns of probability, quantization, and uncertainty. At the level of atoms and particles, reality does not unfold in smooth and continuous ways but through distinct interactions governed by quantum rules.

Yet these discoveries did not make the universe chaotic or random.

On the contrary, quantum theory provided a deeper structure underlying the behavior of matter and energy. It explained the stability of atoms, the formation of chemical bonds, and the properties of materials. Many modern technologies—from transistors and lasers to medical imaging and advanced communication systems—arose directly from this understanding.

Still, the quantum revolution raised new philosophical questions.

What does it mean for reality if particles behave like waves? Why does measurement influence what we observe? How should we interpret a world described by probabilities rather than certainties?

These questions pushed scientists and philosophers to reconsider the nature of observation, knowledge, and reality itself.

As the twentieth century progressed, researchers continued exploring the structure of atoms and the forces governing particles. Their investigations would lead to even deeper insights into the fundamental components of matter and the forces shaping the universe.

In the next chapter, we will explore how scientists uncovered the inner structure of atoms and the particles within them, revealing another layer of reality hidden beneath the visible world.

Reflective Interlude

From Understanding Nature to Shaping It

At this point in our journey, something important has quietly unfolded.

Human thinking has moved through several stages.

First came stories. Early humans looked at the world with wonder and explained natural events through myths, imagination, and symbolic language. These stories helped people live with uncertainty and gave meaning to the forces they could not yet understand.

Then came philosophy. Thinkers began asking deeper questions. Instead of only telling stories, they searched for patterns, causes, and principles behind natural events. The world slowly began to appear more ordered and understandable.

Next came science. Observation, experimentation, and mathematics transformed curiosity into systematic knowledge. Scientists learned how to measure nature, test ideas, and describe physical laws governing motion, light, energy, electricity, and matter.

By this stage, humanity had learned how the universe behaves.

But understanding alone was not the end of the journey.

Once the laws of nature became clear, another transformation began.

Humans started asking a new question:

If we understand these laws, can we use them?

This question gave birth to engineering.

Engineering is the stage where knowledge becomes application. The principles discovered by scientists are translated into machines, technologies, and systems that reshape human life.

Fire became engines. Electricity became power networks. Quantum theory became electronics and computing.

At this point, the relationship between philosophy, science, and technology becomes clear.

Philosophy asks questions. Science discovers the rules. Engineering applies those rules to build the world around us.

In the next part of this book, we will explore how scientific knowledge began transforming into practical systems and technologies—marking the beginning of the engineering age that defines modern civilization.

PART V — ENGINEERING

"Understanding nature is knowledge. Using that understanding to shape the world is engineering."

Chapter 15 — When Knowledge Became Power

Turning scientific understanding into machines and tools

For thousands of years, humans observed nature with curiosity. They watched the movement of the sun and stars, studied the changing seasons, and noticed patterns in wind, fire, water, and living systems. Gradually these observations turned into philosophy, and later into science. Laws of motion, electricity, heat, and light revealed that nature follows consistent principles.

Yet understanding these laws did not immediately transform the world.

A person could know why an apple falls from a tree, but that knowledge alone would not build a bridge. Scientists could describe how electricity behaves, but describing it did not automatically illuminate cities.

Something else was required.

The knowledge had to be applied.

This moment—when understanding begins to shape tools, machines, and systems—is where a new stage of human development begins. Scientific knowledge stops being only an explanation of nature and becomes a means of interacting with it deliberately.

When people learned how heat behaves, they built engines. When they understood electricity, they created power networks and communication systems. When they discovered the properties of materials, they designed machines capable of transforming industries.

At this point, knowledge itself becomes a form of power.

Not power in the sense of domination, but power in the sense of capability—the ability to shape matter, control energy, and design systems that extend human ability far beyond natural limits.

A single machine can perform the work of thousands of people. A communication system can connect entire continents. A small electronic device can process enormous amounts of information.

All of this arises from one fundamental transformation:

Understanding nature becomes the ability to use nature.

This transformation gave birth to the discipline known as engineering.

Engineering stands at the intersection of science and human need. It takes abstract knowledge about physical laws and converts it into structures, machines, and technologies that function in the real world.

Through engineering, humanity learned not only to observe nature—but to work with its principles to build the civilization we live in today.

15.1 — From Curiosity to Control

In the beginning, human interaction with nature was mostly observational.

Early humans watched the world around them and tried to survive within it. They learned from experience—when certain plants grew, how animals behaved, when storms arrived, and where water could be found. This knowledge was practical but limited. Nature appeared powerful and unpredictable, and humans mainly adapted to its conditions.

Curiosity gradually changed this relationship.

People began asking why certain events happened. Why do objects fall? Why does fire produce heat? Why does lightning strike the ground? These questions slowly transformed observation into investigation.

The birth of science marked an important step in this process.

When scientists began studying nature systematically, they discovered that many natural processes follow consistent rules. The motion of objects obeys predictable laws. Heat flows according to measurable principles. Electricity and magnetism interact through identifiable patterns.

These discoveries revealed something remarkable.

Nature is not purely chaotic or mysterious. It operates through structures and relationships that can be understood.

Once these relationships were understood, humans realized that they could go beyond observation.

They could begin to control certain processes.

For example, understanding the behavior of heat allowed engineers to design steam engines that converted thermal energy into mechanical motion. Knowledge of electricity made it possible to generate and transmit electrical power. Understanding the properties of materials led to stronger bridges, more efficient machines, and new methods of manufacturing.

Curiosity therefore evolved into capability.

At first, humans merely observed natural forces such as fire, wind, and water. Later they learned how to guide these forces toward specific purposes. Wind could turn mills. Water could drive turbines. Fire could power engines.

Step by step, the relationship between humans and nature began to shift.

Nature was no longer only something to be endured or admired. It became a system whose principles could be understood and applied.

This shift—from curiosity about nature to the ability to shape it—marks one of the most important turning points in the history of human civilization.

15.2 — Machines as Extensions of Human Ability

Once humans began applying scientific knowledge, tools and machines started evolving beyond simple instruments.

A basic tool such as a hammer or a knife already extends human ability. A hammer allows a person to apply greater force to an object, while a knife allows materials to be cut more efficiently than by hand. These tools amplify what the human body can naturally do.

Machines take this idea much further.

A machine is essentially a system designed to multiply or transform human effort using the principles of physics. Instead of relying only on muscles, machines use controlled energy—heat, electricity, or motion—to perform work.

Consider something as simple as a lever.

A lever allows a small force applied at one end to lift a much heavier object at the other. This principle, known since ancient times, demonstrates how understanding basic mechanical relationships can extend human strength.

Over time, machines became more sophisticated.

Water wheels converted the energy of flowing rivers into mechanical motion that could grind grain or power workshops. Windmills used air currents to perform similar tasks. These early machines allowed communities to perform work continuously without relying solely on human labor.

The development of engines expanded this capability dramatically.

Steam engines converted heat from burning fuel into mechanical movement. Later, internal combustion engines transformed chemical energy from fuels into motion capable of driving vehicles and industrial equipment.

Electric motors introduced another powerful method of converting energy into controlled motion. With electricity, machines could operate more smoothly, more efficiently, and often with greater precision.

Through these developments, machines gradually became extensions of human capability.

A single machine could lift enormous loads, produce thousands of identical products, or move people and goods across vast distances. Tasks that once required large groups of workers could now be performed by mechanical systems operating continuously.

This transformation reshaped human productivity.

Machines did not replace human intelligence, but they greatly multiplied the effect of human effort. Engineers designed systems that allowed natural forces—steam pressure, electrical currents, rotating motion—to perform work far beyond the limits of human strength.

In this way, machines became powerful partners in human activity.

They allowed societies to build cities, manufacture goods, transport resources, and create infrastructure on a scale that would have been impossible through human labor alone.

15.3 — Engineering: The Bridge Between Science and Society

Science reveals how nature works.

It discovers laws, patterns, and relationships governing the behavior of matter, energy, and forces. A scientist may explain how electricity flows through a conductor or how heat moves through a material.

But knowing these principles is only part of the journey.

Someone must translate these ideas into systems that actually function in the real world.

This is where engineering begins.

Engineering acts as a bridge between scientific understanding and human needs. It takes the laws discovered by science and applies them to design machines, structures, and technologies that perform specific tasks.

This process requires a different type of thinking.

A scientist may ask, *What are the fundamental laws governing this phenomenon?* An engineer asks, *How can these laws be used to solve a practical problem?*

For example, once scientists understood electricity and magnetism, engineers developed generators that could produce electrical power and motors that could convert that power into motion. Theoretical knowledge became power plants, lighting systems, and transportation networks.

Engineering therefore involves both creativity and precision.

Engineers must design systems that operate reliably under real-world conditions. Materials must withstand stress, machines must operate safely, and complex systems must function continuously without failure.

This often requires balancing multiple factors—efficiency, cost, safety, durability, and environmental impact.

Engineering also connects knowledge with society.

Bridges allow people to cross rivers and valleys. Water systems deliver clean water to cities. Communication networks connect individuals across continents. Transportation systems move goods and people through complex infrastructures.

Behind each of these systems lies careful engineering design.

In this way, engineering transforms abstract knowledge into structures and technologies that shape everyday life.

It is the discipline where scientific ideas leave the laboratory and become part of the physical world we live in.

15.4 — The Rise of Technological Systems

As engineering advanced, machines and tools gradually evolved into large interconnected systems.

Early technologies often operated in isolation. A water wheel powered a mill beside a river. A steam engine drove machinery inside a factory. Each device performed a specific task within a limited setting.

Over time, however, engineers began designing systems that connected many machines and processes together.

One of the earliest examples was the industrial production system.

Factories combined multiple machines into coordinated workflows. Raw materials entered one side of the factory, passed through several stages of processing, and emerged as finished products. Each machine performed a specific function, and together they formed a continuous production process.

This approach dramatically increased efficiency.

Instead of producing goods individually by hand, industries could manufacture large quantities of identical products with consistent quality. Industrial systems allowed societies to produce textiles, tools, vehicles, and other goods on an unprecedented scale.

Transportation systems soon followed a similar pattern.

Railways connected cities across continents, allowing goods and passengers to travel rapidly over long distances. Later, road networks, shipping routes, and aviation systems created complex global transportation infrastructures.

Electric power systems introduced another major transformation.

Rather than generating electricity only for individual buildings or machines, engineers built power grids capable of distributing electrical energy across entire regions. Large power plants produced electricity that traveled through high-voltage transmission lines to cities, industries, and homes.

These systems allowed energy to be shared across vast networks.

Communication technology also evolved into global systems.

Telegraph networks first connected distant locations through electrical signals traveling along wires. Later, telephone networks, radio communication, satellites, and digital networks formed complex global communication infrastructures.

Through these developments, engineering moved beyond individual machines toward integrated technological ecosystems.

Modern society now relies on multiple interconnected systems operating simultaneously—energy networks, transportation systems, manufacturing processes, and communication infrastructures.

Each system depends on scientific principles and careful engineering design. Together, they support the functioning of modern civilization.

What once began with simple tools gradually evolved into complex technological networks capable of coordinating human activity across the entire planet.

15.5 — The Responsibility of Power

As scientific knowledge turned into engineering capability, humanity gained something unprecedented: the ability to reshape the physical world on a massive scale.

Machines could extract resources from the Earth, generate enormous amounts of energy, move goods across continents, and connect people through global communication networks. Technology amplified human power far beyond what earlier generations could imagine.

But with this power came a new responsibility.

When tools become powerful enough to influence entire societies, their consequences extend far beyond the intentions of their creators. A machine designed to improve efficiency may also affect employment patterns. Industrial production can raise living standards while simultaneously placing pressure on natural resources.

Engineering therefore does not exist in isolation from society.

Every technological system interacts with human communities, economies, and the environment. Decisions about how technologies are designed and used can shape the direction of civilization itself.

Energy systems provide a clear example.

Electric power has transformed modern life, enabling lighting, communication, transportation, and industrial activity. Yet the methods used to generate that energy—whether through fossil fuels, nuclear reactions, or renewable sources—carry different environmental and social consequences.

Similarly, communication technologies have connected people across the world. Information now travels instantly between continents. This connectivity creates opportunities for collaboration and learning, but it also introduces challenges related to information overload, privacy, and social dynamics.

These examples illustrate an important principle.

Technological power is not purely technical. It is also ethical and social.

Engineers and scientists must consider not only whether something can be built, but also how it should be used and what effects it may produce. Responsible innovation requires thinking about safety, sustainability, and the long-term impact of technology on human life.

Throughout history, the relationship between knowledge and responsibility has grown stronger as technological capabilities have increased.

When knowledge becomes power, the question is no longer only *what is possible*. It also becomes *what is wise*.

Closing Transition

The journey from scientific discovery to engineering application represents one of the most powerful transformations in human history.

When humans first began observing nature, they were mostly trying to survive within it. Over time, philosophy helped them ask deeper questions, science helped them discover the laws governing the world, and engineering allowed them to apply those laws in practical ways.

Through this process, knowledge gradually turned into capability.

Steam engines transformed heat into motion and launched the industrial age. Electrical systems illuminated cities and connected continents through communication networks. Advances in materials, chemistry, and mechanics enabled machines capable of building modern infrastructure and transportation systems.

Engineering therefore became the stage where understanding nature translated into shaping the world itself.

Bridges spanning rivers, aircraft crossing oceans, satellites orbiting the Earth, and digital networks linking billions of people all exist because scientific principles were transformed into working systems.

Yet this transformation also reveals something deeper.

Every technological achievement ultimately depends on the same fundamental principles of nature that philosophers once wondered about and scientists later studied. The machines and systems of modern civilization are not separate from the universe—they are expressions of our ability to understand and work with its laws.

As technology continues to advance, humanity faces a new challenge.

The tools we create are becoming increasingly powerful and complex. Artificial intelligence, advanced materials, space exploration, and global communication networks are pushing the boundaries of what technology can achieve.

This raises an important question for the future:

How will humanity balance knowledge, power, and responsibility as technology continues to evolve?

In the next chapter, we will explore how the relationship between science, engineering, and human values shapes the future of civilization.

Chapter 16 — The Age of Technology

Electricity, communication, computing, and modern civilization

Consider an ordinary morning in the modern world.

A person wakes up and turns on a light. Electricity flows through hidden wires inside the walls, instantly illuminating the room. A smartphone is checked for messages that may have traveled thousands of kilometers through invisible communication networks during the night.

Outside, transportation systems are already moving millions of people through cities—trains running on electric power, vehicles navigating complex road networks, aircraft crossing continents guided by satellites.

Inside homes and workplaces, computers process enormous amounts of information. Machines assist in manufacturing goods, medical devices monitor health, and digital networks connect individuals across the globe.

What once seemed extraordinary has become routine.

Most people rarely pause to think about the vast technological systems operating quietly around them. Electricity, communication networks, and computing infrastructure function continuously, forming an invisible foundation supporting modern life.

Yet these technologies did not appear suddenly.

They emerged gradually from centuries of scientific discovery and engineering development. Understanding electricity allowed engineers to build power systems. Discoveries in electromagnetism led to communication technologies. Advances in quantum physics and materials science eventually made modern electronics and computing possible.

Together, these innovations created what we now call the technological age.

In this age, human activity is deeply intertwined with machines and systems designed through engineering. Energy flows through power grids, information moves through digital networks, and automated systems assist in everything from transportation to healthcare.

Technology has expanded human capability to an extraordinary degree.

It allows people to communicate instantly across continents, access vast stores of knowledge, and build systems that support billions of individuals living in complex societies.

Yet the technological age also represents something deeper than the presence of machines.

It reflects humanity's ability to translate knowledge into systems that reshape civilization itself.

16.1 — Electricity as the Backbone of Modern Life

If there is one force that quietly sustains the modern technological world, it is electricity.

Unlike many other forms of energy, electricity can be transmitted easily, converted into many different forms, and controlled with great precision. Because of these properties, electricity became the foundation upon which most modern technologies operate.

In earlier centuries, energy sources were limited and localized. Water wheels depended on nearby rivers. Windmills required favorable wind conditions. Steam engines relied on burning fuel at the location where mechanical work was needed.

Electricity changed this relationship between energy and location.

Through power plants and transmission systems, electrical energy could be generated at one place and delivered to another through networks of wires and cables. Large-scale power stations—using coal, water turbines, nuclear reactions, or renewable sources—produce electricity that travels through transmission lines to cities, industries, and homes.

This ability to distribute energy across long distances created entirely new possibilities.

Factories could operate continuously with electrically powered machinery. Homes could be illuminated with electric lighting far more safely and efficiently than candles or gas lamps. Hospitals could operate sophisticated medical equipment requiring stable electrical supply.

Electric motors became one of the most widely used applications of electricity.

These motors convert electrical energy into mechanical motion and are found in countless devices—fans, pumps, elevators, washing machines, industrial equipment, and transportation systems. Electric motors are often compact, efficient, and capable of precise control.

Electricity also made modern lighting possible.

Electric lamps replaced older lighting systems that relied on burning fuels. Streets, buildings, and workplaces could now remain illuminated long after sunset, extending human activity into the night.

Beyond lighting and machinery, electricity powers a vast array of devices that shape everyday life.

Heating and cooling systems regulate indoor environments. Refrigeration preserves food and medicines. Medical equipment supports diagnosis and treatment in healthcare facilities.

Because electricity can be converted easily into light, motion, heat, or information signals, it has become one of the most versatile energy forms ever used by humanity.

In modern civilization, electrical networks function like a circulatory system, distributing energy throughout the infrastructure that supports daily life.

Without this invisible flow of electrical energy, most technological systems that define modern society would simply cease to function.

16.2 — The Communication Revolution

While electricity provided the energy that powered machines and cities, it also made something even more transformative possible—the rapid transmission of information.

For most of human history, communication moved at the speed of physical travel. Messages had to be carried by messengers, ships, or animals across long distances. Important news or instructions could take days, weeks, or even months to reach their destination.

This limitation shaped the structure of societies.

Governments, trade networks, and communities were constrained by how quickly information could move. Decisions made in one location often arrived too late to influence events happening elsewhere.

Electricity changed this relationship dramatically.

The first major breakthrough came with the invention of the telegraph. By sending electrical signals through wires, coded messages could travel across long distances almost instantly. Information that once required days of travel could now be transmitted within minutes.

The telegraph transformed commerce, journalism, and government coordination. Financial markets, railway systems, and international communication began operating on much faster timescales.

Soon after, the development of the telephone allowed people to transmit not just coded signals but human voices. Conversations could occur in real time between individuals separated by large distances. This innovation fundamentally changed personal and business communication.

As technology advanced further, communication systems expanded beyond wires.

Radio waves allowed messages to travel through the air without physical connections. Broadcasting systems enabled information and entertainment to reach large audiences simultaneously. Later, television introduced visual communication across long distances.

The next major transformation arrived with digital communication and the internet.

Digital systems convert information into electrical signals that can be transmitted through networks of computers and communication devices. Through these networks, text, images, sound, and video can move across the world almost instantly.

Satellites orbiting Earth further extended the reach of communication systems. Signals can now travel across continents and oceans, connecting remote regions with global information networks.

Today, billions of people communicate through interconnected digital platforms. Messages, documents, and media circulate continuously through complex communication infrastructures linking individuals, organizations, and entire societies.

The communication revolution therefore reshaped how humans interact with one another.

Information that once moved slowly and locally now travels at extraordinary speed across a globally connected network. This transformation has changed commerce, education, governance, and social relationships in profound ways.

Electricity did not only power machines—it created the possibility for instant global communication, bringing humanity closer together than ever before.

16.3 — The Rise of Computing

As communication systems expanded and information began flowing rapidly through electrical networks, another technological transformation quietly began.

Humans started building machines capable of processing information itself.

For centuries, calculations had been performed manually using pen and paper, abacuses, or mechanical devices. These methods were sufficient for many tasks, but as scientific research, engineering projects, and industrial systems became more complex, the need for faster and more reliable calculation grew.

Early mechanical calculating machines were developed to assist with arithmetic operations. These devices could add, subtract, and sometimes multiply numbers more efficiently than manual methods.

However, the real breakthrough came when engineers began using electronic circuits to perform calculations.

Electronic components can switch electrical signals on and off extremely quickly. By organizing these signals into logical patterns, engineers discovered that machines could perform sequences of operations automatically.

This principle led to the development of the electronic computer.

One of the pioneers of modern computing theory was Alan Turing. Turing showed that machines could be designed to follow logical instructions and process information step by step, much like a human performing calculations.

Early computers were large and complex machines occupying entire rooms. They were primarily used for scientific research, military calculations, and engineering analysis.

Over time, advances in electronics allowed computers to become smaller, faster, and more efficient.

The invention of the transistor replaced bulky vacuum tubes with tiny semiconductor devices capable of controlling electrical signals with great precision. Later, integrated circuits allowed thousands—and eventually billions—of electronic components to be placed on a single chip.

These developments made modern computing possible.

Today, computers exist in many forms: powerful servers managing global data networks, personal computers used for work and education, and small embedded processors operating inside everyday devices.

Computers process enormous quantities of information, supporting activities ranging from scientific simulations and financial transactions to navigation systems and digital communication.

Through computing, technology moved beyond controlling physical energy and motion.

Machines began assisting humans in handling information and knowledge itself.

This shift has had a profound impact on modern civilization, shaping how people work, communicate, learn, and make decisions in an increasingly data-driven world.

16.4 — A Connected Global Civilization

As electricity, communication systems, and computing technologies matured, they began forming something much larger than individual inventions.

They created a globally interconnected technological civilization.

In earlier eras, societies often functioned as relatively isolated regions. Trade routes connected distant cultures, but communication and transportation were slow. Economic activity, knowledge, and governance operated mainly within local or regional boundaries.

Modern technology transformed this situation.

Electric power networks enabled industries to operate at large scales. Transportation systems—railways, shipping routes, highways, and air travel—allowed goods and people to move rapidly across continents. Communication technologies connected individuals across nations and cultures.

Together, these systems created a world where activities in one region can quickly influence events elsewhere.

For example, financial markets in different countries respond almost instantly to economic developments occurring across the globe. Scientific discoveries made in one laboratory can be shared with researchers worldwide within minutes through digital networks.

Global communication platforms allow people to exchange ideas, collaborate on projects, and participate in shared discussions regardless of geographic distance.

The internet has become one of the most powerful examples of this interconnected system.

Through the internet, billions of devices are linked together in a vast network that carries information continuously across the planet. Educational resources, research data, news, entertainment, and communication all travel through this digital infrastructure.

Transportation systems also reflect this global connectivity.

Air travel enables people to cross oceans within hours. Shipping networks move raw materials and manufactured goods through complex supply chains connecting multiple countries.

Manufacturing itself has become part of global production systems.

Components designed in one country may be manufactured in another, assembled in a third location, and distributed worldwide. Modern supply chains depend on precise coordination enabled by communication networks and digital information systems.

These developments have reshaped the structure of human civilization.

Economies, cultures, and institutions are increasingly linked through technological networks. Knowledge spreads more rapidly, collaboration becomes easier across distances, and global challenges can be addressed through coordinated international efforts.

The technological systems created through engineering have therefore done more than improve efficiency.

They have connected humanity into a shared global civilization, where information, resources, and ideas circulate continuously across the world.

16.5 — Living in a Technological Environment

As technological systems expanded and interconnected, something subtle but profound happened.

Humans no longer simply *use* technology.

We now live inside technological environments.

In earlier centuries, tools were mostly external objects that people used occasionally—a plough in a field, a hammer in a workshop, a lantern for light at night. These tools assisted specific tasks but did not surround daily life continuously.

Modern technology is different.

Cities operate through networks of electricity, water supply, transportation systems, communication infrastructure, and digital services. Buildings contain heating, cooling, and lighting systems controlled through electrical networks. Vehicles rely on complex engineering systems guided by digital navigation and control technologies.

Many aspects of daily life now depend on technological systems functioning smoothly in the background.

When electricity flows normally, lights turn on, machines operate, communication networks function, and transportation systems run according to schedule. When these systems stop, even briefly, their importance becomes immediately visible.

Technology also shapes how people work and learn.

Digital tools allow individuals to access enormous stores of information, collaborate across distances, and perform tasks that once required specialized equipment or large teams. Education increasingly involves digital platforms, simulations, and global knowledge resources.

Healthcare has also been transformed.

Medical technologies allow doctors to monitor the human body with remarkable precision. Imaging systems reveal internal structures, electronic devices assist in diagnosis and treatment, and advanced surgical tools improve the safety and effectiveness of medical procedures.

Even social interactions are influenced by technological environments.

People communicate through messaging platforms, share ideas through online networks, and participate in global conversations made possible by digital infrastructure.

These developments illustrate how deeply technology has become integrated into human life.

The modern human experience is no longer defined solely by natural surroundings but also by engineered systems that shape the environment in which we live.

Electricity, communication networks, computing systems, and transportation infrastructures form a technological ecosystem supporting modern civilization.

Understanding this environment is therefore not only a matter of technical knowledge—it is also part of understanding the world we inhabit today.

Closing Transition

The technological age represents the culmination of a long journey.

What began with curiosity about the stars, fire, and storms gradually evolved into philosophy, then science, and finally engineering. Step by step, humanity learned how nature behaves and how those principles could be used to build tools, machines, and complex technological systems.

Today, those systems form the foundation of modern civilization.

Electric power flows through vast networks supporting cities and industries. Communication technologies allow people to exchange information instantly across continents. Computing systems process enormous quantities of data, guiding decisions in science, medicine, economics, and governance.

Through these developments, technology has extended human capability in extraordinary ways.

It allows societies to produce food and goods at large scales, connect billions of people through shared information networks, and explore environments ranging from the depths of the oceans to the far reaches of space.

Yet this technological transformation also brings a deeper realization.

Technology itself is not separate from the human mind that created it. Every machine, system, and network reflects the knowledge, values, and decisions of the people who design and use it.

In other words, technology is not only about devices or infrastructure.

It is also about how humans understand and interact with the world.

As technological systems grow more powerful and complex, this relationship becomes increasingly important. The future of civilization will depend not only on what technologies we develop, but also on how wisely we choose to apply them.

In the final part of this book, we will step back from machines and systems to explore a deeper question:

What does this long journey—from stories to science to technology—reveal about the human mind and its role in shaping the future?

Chapter 17 — Engineers: The Translators of Nature

How engineering converts natural laws into practical solutions

Nature already contains the rules that govern how the universe works.

The motion of planets follows the laws of gravity. Heat flows according to the principles of thermodynamics. Electricity moves through conductors according to the laws of electromagnetism. Light travels as electromagnetic waves, and matter behaves according to the principles of atomic and quantum physics.

These laws existed long before humans discovered them.

Scientists devote their efforts to understanding these natural principles. Through observation, experiment, and reasoning, they uncover the patterns that describe how nature behaves.

But understanding alone does not build bridges, power cities, or create communication networks.

Something more is required.

The knowledge must be translated into practical systems that function reliably in the real world.

This is where engineers enter the story.

Engineers take the language of science—the equations, principles, and physical laws—and convert them into machines, structures, and technologies that solve real problems. They interpret what nature allows and design systems that work within those boundaries.

In this sense, engineers act as translators of nature.

Just as a translator converts ideas from one language into another, engineers convert scientific understanding into practical designs. The laws of physics become engines. The principles of electromagnetism become power grids and communication networks. Knowledge of materials becomes buildings, aircraft, and medical devices.

Through this process, abstract knowledge becomes part of everyday life.

When a bridge carries traffic across a river, when an airplane crosses an ocean, or when a medical device saves a patient's life, the invisible laws of nature are quietly at work within engineered systems.

Engineering therefore represents one of the most powerful expressions of human creativity.

It is the discipline where understanding the universe becomes the ability to shape the world around us.

17.1 — Reading the Language of Nature

Before an engineer can design a machine, build a bridge, or create a communication system, something fundamental must happen.

The engineer must first understand the language of nature.

Nature does not communicate in human words. It expresses itself through relationships between forces, motion, energy, materials, and fields. These relationships appear as patterns that can be described through scientific laws.

For example, gravity determines how structures bear weight. Thermodynamics governs how heat flows through materials and engines. Electromagnetism explains how electrical signals travel through circuits and communication networks. The behavior of materials depends on atomic structures and chemical interactions.

These laws form the vocabulary of engineering.

An engineer studies these principles carefully because every design must operate within the limits defined by nature. A bridge must withstand gravitational forces and environmental stress. An aircraft must obey the principles of aerodynamics to generate lift. An electronic circuit must follow the rules governing electrical currents and resistance.

In this sense, engineering begins with listening to nature.

The natural world already defines what is possible and what is not. Engineers do not invent new physical laws; instead, they interpret existing ones and learn how to work with them.

This requires a deep understanding of multiple scientific disciplines.

Mechanical engineers study motion, forces, and materials. Electrical engineers analyze circuits, electromagnetic fields, and signal transmission. Chemical engineers examine reactions, fluid behavior, and energy transfer. Civil engineers study structures, soils, and environmental forces.

Although these fields focus on different systems, they all share the same foundation: understanding how natural laws operate in real conditions.

When engineers design something, they are essentially asking a question:

Given the laws of nature, how can we arrange materials and energy so that a system performs a desired function?

This process requires careful analysis and precise calculations. But it also requires intuition—an ability to visualize how forces interact and how systems behave under changing conditions.

By learning to read the language of nature, engineers gain the ability to design systems that cooperate with physical laws rather than struggle against them.

In this way, engineering begins not with machines or structures, but with understanding the principles that govern the universe itself.

17.2 — Turning Principles into Designs

Once engineers understand the scientific principles governing a phenomenon, the next step begins.

They must convert those principles into designs that perform a specific function.

This transformation—from principle to design—is where engineering truly takes shape.

Scientific laws describe how nature behaves under certain conditions. For example, the laws of motion explain how forces influence movement, while the laws of electricity describe how currents flow through conductors. These laws provide a framework for understanding natural behavior.

However, a scientific law by itself does not specify how to build a machine.

Engineering begins when someone asks a practical question:

How can this principle be used to solve a real-world problem?

For instance, the understanding of steam expansion in thermodynamics led engineers to design steam engines capable of converting heat into mechanical motion. Knowledge of electromagnetism allowed engineers to create electric motors and generators. Principles of aerodynamics guided the development of aircraft capable of sustained flight.

In each case, engineers translated abstract principles into concrete systems.

This process requires several steps.

First, the engineer identifies the objective—the function the system must perform. A bridge must carry vehicles across a river. A pump must move water from one location to another. A communication system must transmit information reliably across distance.

Next comes the design phase.

Engineers determine how materials, energy sources, and physical structures should be arranged to achieve the desired function. This stage often involves calculations, modeling, and testing to ensure that the system will behave as expected.

For example, when designing a bridge, engineers analyze loads, structural forces, and material strengths. When developing an electrical device, they examine current flow, resistance, and power requirements.

Finally, the design must be implemented.

Materials are selected, components are manufactured, and the system is assembled according to the planned structure. Once built, the design is tested under real conditions to verify that it performs safely and reliably.

This entire process illustrates how engineering transforms scientific knowledge into practical solutions.

Where science explains what happens, engineering determines how to make it useful.

Through this translation, natural laws become tools that allow humans to build machines, structures, and technologies capable of solving complex problems.

17.3 — The Balance Between Theory and Reality

Scientific laws often describe nature in precise and elegant forms.

Equations express relationships between forces, energy, motion, and materials with mathematical clarity. These models allow scientists to predict how systems should behave under ideal conditions.

But the real world rarely behaves in perfectly ideal ways.

Materials contain imperfections. Environmental conditions change. Machines experience wear and stress. Systems must operate safely even when unexpected events occur.

Engineering exists precisely at this intersection between ideal theory and practical reality.

An engineer cannot rely only on theoretical equations. The design must also account for real-world factors such as material limitations, environmental forces, manufacturing tolerances, and long-term reliability.

Consider the example of a bridge.

Physics can describe how loads distribute across structural components. However, an engineer must also consider wind forces, temperature changes, corrosion, vibrations from traffic, and the possibility of earthquakes. Safety margins must be included to ensure the structure remains stable even under extreme conditions.

The same principle applies to many engineering systems.

In electronics, circuits must tolerate fluctuations in voltage and temperature. In aircraft design, engineers must account for turbulence, structural fatigue, and variations in atmospheric conditions. In power systems, electrical networks must remain stable even when demand changes suddenly.

Because of these uncertainties, engineering design often includes factors of safety.

These safety margins ensure that structures and systems can withstand conditions beyond normal operating limits. Instead of designing for only the expected load or stress, engineers prepare systems to handle unexpected situations.

Engineering therefore involves continuous testing and refinement.

Designs are analyzed through simulations, laboratory experiments, and real-world trials. Engineers study how systems behave under different conditions and adjust their designs to improve performance and reliability.

This process highlights an important truth.

Scientific theory provides the foundation, but engineering must work with the complexity of reality.

The success of an engineered system depends not only on understanding physical laws but also on anticipating how those laws interact with materials, environments, and human use.

In this way, engineering becomes a discipline of balance—where precise scientific knowledge meets the practical challenges of the real world.

17.4 — Engineering as Creative Problem Solving

At first glance, engineering may appear to be a purely technical discipline based on calculations and scientific principles.

But in practice, engineering is also a deeply creative activity.

When engineers face a problem, there is rarely only one possible solution. Many different designs may satisfy the same scientific laws, yet each design may vary in efficiency, cost, safety, and practicality.

The engineer must therefore imagine new ways to arrange materials, forces, and energy so that a system performs its intended function.

This process often begins with identifying the core problem.

For example, how can a structure span a wide river safely? How can a machine convert energy more efficiently? How can information travel reliably across long distances? Each of these questions requires understanding both the physical laws involved and the practical constraints of the situation.

From there, engineers begin exploring possible solutions.

They may sketch designs, create models, or simulate systems using computational tools. Different configurations are analyzed to determine how forces will act within the system and how components will interact.

This stage requires both analytical thinking and imagination.

An engineer must visualize how parts will behave together, often predicting outcomes before a system is actually built. Sometimes a small change in structure or arrangement can dramatically improve the performance of a design.

Throughout history, many engineering breakthroughs have emerged from this creative problem-solving process.

The development of suspension bridges allowed structures to span enormous distances using tension instead of compression. The design of jet engines enabled aircraft to travel faster and higher than earlier propulsion systems. Modern microelectronics compress billions of transistors into tiny semiconductor chips capable of performing complex computations.

Each of these innovations required engineers to look beyond existing designs and imagine new possibilities.

Creativity in engineering does not mean ignoring scientific laws.

Instead, it involves working within those laws in innovative ways. Engineers explore different configurations and approaches until they find solutions that satisfy both the requirements of physics and the needs of society.

In this sense, engineering resembles a form of structured creativity.

Scientific principles provide the boundaries within which designs must operate, but within those boundaries there remains enormous space for innovation and discovery.

Through this creative process, engineers continually develop new systems that expand human capability and transform the world around us.

17.5 — The Engineer's Responsibility to Society

Engineering does not exist only in laboratories or technical drawings.

The systems engineers design become part of everyday life. Bridges carry millions of people across rivers and valleys. Power grids deliver electricity to cities. Communication networks connect individuals across continents. Medical devices support human health.

Because of this direct influence, engineering carries a significant responsibility toward society.

Unlike many theoretical disciplines, the results of engineering decisions affect public safety, economic systems, environmental conditions, and human well-being. A poorly designed structure can collapse. A malfunctioning medical device can endanger patients. A flawed infrastructure system can disrupt entire communities.

For this reason, engineers must consider more than technical performance.

They must think about safety, reliability, and long-term consequences. Designs are evaluated carefully to ensure that systems operate within safe limits and remain stable under unexpected conditions. Regulations, standards, and testing procedures exist to help protect the public from potential risks.

Environmental responsibility has also become an important aspect of engineering.

Large-scale industrial systems can influence natural ecosystems, energy consumption, and resource availability. Engineers increasingly work to design technologies that reduce pollution, improve efficiency, and support sustainable development.

Another responsibility involves ethical decision-making.

Engineering projects often involve choices that affect large groups of people. Decisions about transportation systems, energy infrastructure, or digital technologies can influence how societies function. Engineers must therefore balance technical possibilities with ethical considerations about fairness, access, and social impact.

At the same time, engineering has the potential to improve lives in meaningful ways.

Clean water systems prevent disease. Reliable energy networks support healthcare, education, and economic activity. Communication technologies connect communities and expand access to knowledge.

These achievements remind us that engineering is not only about machines and structures.

It is ultimately about serving human needs.

The knowledge used to build technologies carries the power to shape the future of civilization. With that power comes the responsibility to ensure that technological progress contributes to safety, well-being, and the sustainable development of human society.

Closing Transition

As we step back and look at the entire journey traced through this book, a remarkable pattern becomes visible.

Human understanding did not emerge suddenly. It evolved gradually through several stages.

First came stories. Early humans tried to make sense of nature through myths and imagination. These stories were not scientific explanations, but they represented the first attempts to interpret the world.

Then came philosophy. Thinkers began asking deeper questions about causes, patterns, and principles governing the universe. Instead of accepting events as mysterious or random, they searched for order and meaning.

Next came science. Observation, experimentation, and mathematics revealed the laws governing motion, energy, light, electricity, and matter. Nature began to appear as a system operating according to consistent principles.

From there emerged engineering.

Engineers translated scientific knowledge into machines, structures, and systems that support human civilization. Power grids, communication networks, transportation systems, and digital technologies all arose from this ability to apply natural laws in practical ways.

Seen from this perspective, engineering represents the moment when knowledge becomes capability.

Understanding the universe is no longer only about explanation. It becomes the ability to design, build, and transform the environment in which we live.

Yet this journey also reveals something deeper.

The same curiosity that once led early humans to watch the stars and wonder about fire still drives modern scientific and technological progress. Behind every machine and every technological system lies the same human impulse—to understand nature and to work with its principles.

In the final chapter of this book, we will reflect on the larger meaning of this journey.

From stories around ancient fires to advanced technologies shaping modern civilization, the path of human knowledge reveals how philosophy, science, and engineering are all part of a single unfolding process: humanity learning how to understand and interact with the universe.

Reflective Interlude

The Journey So Far

Before we enter the final part of this book, it is worth pausing for a moment and looking back at the path we have traveled.

The story we followed is not only the history of science or technology. It is the story of *how the human mind learned to understand the universe.*

At the beginning, humans lived close to nature with very little knowledge of the forces surrounding them. Fire, storms, the movement of the stars, birth and death—all appeared mysterious. To make sense of these experiences, people created stories and myths. These stories were the first language through which humans tried to interpret the world.

Gradually, curiosity deepened.

Thinkers began asking questions about causes and patterns. Philosophy emerged as an attempt to understand the order hidden behind natural events. The universe was no longer explained only through stories but through ideas about reason, harmony, and structure.

With time, observation and experimentation transformed curiosity into science.

Scientists learned to measure nature, test ideas, and describe the laws governing motion, heat, electricity, light, and matter. The universe began to reveal itself as a system governed by consistent principles.

Engineering then carried this understanding into the physical world.

The knowledge discovered by science became machines, infrastructure, communication systems, and technologies that define modern civilization. The laws of nature were no longer only understood—they were applied to build the world in which we now live.

Through this long process, something remarkable happened.

Humanity moved from *explaining nature* to *working with nature*.

But the journey does not end with machines and technology.

In fact, this is where a deeper transition begins.

As knowledge and technology expand, humans must also reflect on their own role within this evolving system. The tools we create influence how we think, how we interact with one another, and how we shape the future of civilization.

The final part of this book therefore shifts the focus.

Instead of examining the development of knowledge and technology, we will explore the *human transition*—how our understanding of the universe, and our growing technological power, reshape the way we live, think, and imagine the future.

PART VI — THE HUMAN TRANSITION

"The greatest discovery humanity ever made was not a machine or a theory — it was the realization that we can understand the universe."

Chapter 18 — The Curious Mind

Why every scientist, engineer, and philosopher begins as a curious human

Long before a person becomes a scientist, an engineer, or a philosopher, something much simpler appears first.

Curiosity.

Almost every child begins life by asking questions. The world appears new, mysterious, and full of possibilities. A child looks at the sky and wonders why it is blue. They drop a toy and ask why it falls to the ground. They look at the stars at night and ask where they go during the day.

These questions may sound simple, even naïve.

Yet hidden inside them are the same questions that eventually shaped the history of human knowledge.

Why do objects fall? Why do the planets move across the sky? What is light? What is matter? How does electricity travel through wires?

At the beginning, these questions do not come from textbooks or laboratories. They arise from a natural human impulse to understand the world.

Curiosity does not require specialized knowledge. It only requires the willingness to notice something and ask *why*.

Many of the greatest discoveries in science began exactly this way.

A falling apple raised questions about gravity. Strange sparks produced by rubbing materials together led to the study of electricity. Observing the colors of heated objects helped reveal the laws of radiation and quantum physics.

Behind each of these discoveries was not only intelligence or technical skill, but also a curious mind that refused to ignore simple questions.

In fact, curiosity is often the most important starting point for learning.

Without curiosity, knowledge becomes something memorized but not truly understood. With curiosity, even ordinary experiences become opportunities for discovery.

A scientist continues asking questions about nature. An engineer asks how natural principles can be used to solve problems. A philosopher asks deeper questions about meaning, existence, and understanding.

But all of them begin from the same place.

They begin as curious human beings trying to understand the world around them.

18.1 — Curiosity as the Starting Point of Knowledge

Every major discovery in human history begins with something surprisingly simple.

A question.

Curiosity appears when a person notices something that does not immediately make sense. An ordinary event—an object falling, a shadow moving, a spark appearing—suddenly becomes a source of wonder. Instead of accepting it as routine, the curious mind asks why it happens.

This simple act marks the beginning of knowledge.

Before scientific instruments existed, early humans observed the natural world carefully. They watched the changing phases of the moon, the patterns of the seasons, and the behavior of animals and plants. These observations gradually formed the earliest foundations of understanding.

Over time, curiosity turned observation into investigation.

People began asking questions such as: Why do rivers always flow downhill? Why do certain materials burn while others do not? Why do stars appear to move across the sky?

Each question opened a path toward deeper understanding.

For example, curiosity about the motion of celestial bodies eventually led to the development of astronomy. Questions about falling objects contributed to the discovery of the laws of motion and gravity. Investigations into heat and fire revealed the principles of thermodynamics.

In every case, the process begins the same way.

Someone notices something and refuses to ignore the mystery behind it.

Curiosity transforms ordinary experiences into opportunities for discovery. It encourages people to look more closely, test ideas, and seek explanations that reveal patterns within nature.

This process does not belong only to professional scientists.

Curiosity is a natural human ability. Farmers observing weather patterns, craftsmen experimenting with materials, and explorers studying unfamiliar environments all contribute to expanding knowledge through careful observation and questioning.

Over generations, these individual questions accumulate.

What begins as simple curiosity eventually grows into organized fields of knowledge—science, engineering, medicine, and many others.

In this sense, curiosity is not merely a personality trait.

It is the starting point of human understanding.

18.2 — The Courage to Question

Curiosity may begin with a question, but progress requires something more.

It requires the courage to question what everyone else accepts.

Throughout history, many ideas about the world were widely believed for long periods of time. People assumed that the Earth stood still at the center of the universe, that heavier objects fall faster than lighter ones, or that heat was a mysterious substance flowing through materials.

These explanations often appeared reasonable because they matched everyday experience or long-standing traditions.

Yet scientific progress occurs when someone dares to ask:

What if this explanation is incomplete?

Questioning accepted ideas is not always easy.

Established beliefs are often supported by tradition, authority, or social consensus. When someone challenges them, the idea may initially seem strange or even incorrect. Many discoveries in science began as unconventional suggestions that required careful evidence before they were accepted.

Consider the idea that the Earth moves around the Sun.

For centuries, people believed the opposite because the ground beneath their feet seemed perfectly still. It required careful observation and the courage to challenge tradition before a new understanding of the solar system emerged.

The same pattern appears in many areas of science.

New theories often begin as attempts to explain observations that existing ideas cannot fully account for. At first, these proposals may appear unusual. But if evidence continues to support them, they gradually reshape our understanding of nature.

The courage to question does not mean rejecting knowledge.

Instead, it means examining ideas carefully and being willing to improve them when new evidence appears.

Scientists, engineers, and philosophers all rely on this attitude. Progress becomes possible when individuals remain open to new insights while testing them carefully against observation and reason.

In this way, curiosity evolves into inquiry.

The simple act of asking "why" becomes a powerful tool for exploring the universe and expanding human knowledge.

18.3 — Learning Through Observation

Curiosity encourages people to ask questions, but observation provides the clues needed to answer them.

Before complex instruments and laboratories existed, the natural world itself served as the first classroom. Early humans learned by watching carefully—observing how plants grew, how animals behaved, how rivers flowed, and how the sky changed throughout the day and night.

Through repeated observation, patterns gradually became visible.

The sun rises in the east and sets in the west. The seasons follow a regular cycle. Certain stars appear only during particular times of the year. These patterns suggested that nature operates according to consistent rules rather than random events.

Observation therefore became the foundation of scientific thinking.

A curious mind does not only ask questions—it also looks closely at what is happening. Small details that others might ignore can reveal important relationships within nature.

For example, careful observation of falling objects helped scientists understand the laws of motion. Observing how light reflects and bends led to the study of optics. Watching electrical sparks jump between materials eventually contributed to the discovery of electric charge.

Even complex discoveries often begin with simple observations.

A scientist may notice an unexpected result in an experiment. An engineer may observe a machine behaving differently under certain conditions. These observations create opportunities for deeper investigation.

Observation also teaches patience.

Nature rarely reveals its patterns immediately. Many discoveries required years of careful measurement and repeated experiments before reliable conclusions could be drawn.

This process shows that knowledge grows gradually.

Curiosity asks the question. Observation gathers the evidence. Together they form the beginning of understanding.

In this way, learning through observation transforms everyday experiences into steps toward discovering the deeper principles that govern the natural world.

18.4 — Curiosity and Creativity

Curiosity encourages us to ask questions and observe the world carefully, but another human ability plays an equally important role in discovery.

That ability is creativity.

Creativity allows the mind to imagine possibilities beyond what is immediately visible. When a curious person encounters a mystery, creativity helps them explore different ways the phenomenon might be explained.

In science and engineering, this process often begins with forming a hypothesis—an idea about how something might work.

For example, early scientists trying to understand the motion of planets had to imagine models of the solar system before they could test them through observation. Engineers designing new machines often visualize how different components might interact before building the actual system.

Creativity therefore helps bridge the gap between observation and explanation.

A curious observer may notice that certain patterns exist in nature, but creative thinking allows them to propose new ways of interpreting those patterns.

Many important scientific ideas first appeared as imaginative possibilities.

The idea that invisible forces such as gravity act across space required imaginative thinking. The concept that electricity flows through conductors like a current required envisioning unseen processes. Even modern ideas about atoms, fields, and quantum behavior began as attempts to imagine what might exist beyond direct observation.

Creativity also plays an essential role in engineering.

When engineers face a practical challenge—such as designing a more efficient engine or building a structure capable of spanning a wide distance—they must imagine solutions that do not yet exist. By exploring different configurations of materials, forces, and energy, they create systems capable of solving complex problems.

This creative process does not ignore scientific laws.

Instead, creativity works within the boundaries of natural principles, exploring different ways those principles can be arranged and applied.

Through curiosity and creativity working together, humans have been able to develop new ideas, discover natural laws, and build technologies that reshape civilization.

In this way, creativity transforms questions about the world into new ways of understanding and interacting with it.

18.5 — Keeping Curiosity Alive

Curiosity is natural in early childhood, but maintaining it throughout life requires effort.

Children often ask many questions because they are encountering the world for the first time. Every experience appears new and mysterious. As people grow older, however, routines and established knowledge sometimes reduce the impulse to question familiar things.

Education plays an important role in shaping this process.

When learning encourages memorization without exploration, curiosity may gradually weaken. Students may focus on repeating answers rather than understanding the deeper reasons behind them. Over time, knowledge can become something that feels fixed rather than something open to discovery.

But the most meaningful learning happens when curiosity remains active.

When students ask why a principle works instead of only accepting it, they begin to connect ideas and develop deeper understanding. When they experiment, observe, and test their own ideas, learning becomes a process of discovery rather than simple repetition.

Teachers who encourage curiosity often create environments where questions are welcomed.

Instead of treating uncertainty as a problem, they treat it as an opportunity. A question that does not yet have an answer can lead to exploration, investigation, and new insights.

Curiosity also requires freedom from the fear of making mistakes.

In science and engineering, many attempts do not immediately succeed. Experiments may produce unexpected results, and designs may need several revisions before they work properly. These experiences are not failures but steps in the learning process.

When curiosity remains active, people continue exploring new ideas throughout their lives.

Scientists investigate unanswered questions about nature. Engineers design new technologies to solve emerging problems. Philosophers examine the deeper meaning of knowledge and existence.

In each case, curiosity keeps the mind open to discovery.

Maintaining curiosity therefore becomes one of the most valuable habits for intellectual growth. It allows individuals to continue learning, adapting, and contributing to the expansion of human understanding.

Closing Transition

The curious mind is the quiet beginning of every intellectual journey.

Before there are equations, laboratories, machines, or technologies, there is simply a person asking a question about the world. A falling object, a moving shadow, a spark of electricity, or a distant star becomes the starting point for exploration.

From such simple moments, entire fields of knowledge have emerged.

Curiosity leads to observation. Observation leads to questions. Questions lead to investigation, and investigation gradually reveals the principles that govern nature. Over time, these principles become the foundation for science and engineering.

But curiosity does something even more important.

It keeps the human mind open.

A curious person is willing to look again at familiar things and ask whether there might be a deeper explanation. This openness allows knowledge to grow and evolve rather than remain fixed.

Throughout history, every scientist, engineer, and philosopher has followed this same path. They began not as experts, but as individuals who refused to ignore the mysteries around them.

In this sense, curiosity is not only the beginning of learning—it is also the force that keeps human knowledge moving forward.

In the final chapter of this book, we will reflect on the larger meaning of this journey. From the earliest stories told around ancient fires to the technologies shaping modern civilization, the evolution of knowledge reveals how the human mind continues to explore, understand, and reshape the world.

Chapter 19 — Two Languages of Understanding

Intuition and equations — why both are necessary to understand reality

Imagine a classroom where a teacher is explaining a scientific idea. One student understands the concept when the teacher draws a simple diagram on the board and describes how the system behaves. Another student begins to understand only when the same idea is written in the form of equations and relationships between quantities.

Both students are learning about the same phenomenon.

But they are using different languages of understanding.

Human beings often grasp ideas first through intuition. We imagine how something behaves, visualize motion or patterns, and relate new ideas to experiences we already understand. When someone explains why an object falls or how light reflects from a mirror using examples and images, the mind builds a mental picture of the process.

This intuitive language helps people see the logic behind natural phenomena.

Mathematics, however, provides a different kind of description.

Instead of pictures and examples, it uses symbols, equations, and precise relationships. Mathematical expressions allow scientists to measure quantities, calculate outcomes, and predict how systems will behave under different conditions.

At first, these two approaches may seem very different.

Intuition feels natural and accessible, while mathematics often appears abstract and formal. Yet both describe the same underlying reality.

In fact, throughout the history of science, the most powerful insights have emerged when these two languages work together.

An intuitive idea suggests how a system might behave. Mathematics then provides a precise way to test and refine that idea. When both perspectives align, understanding becomes deeper and more reliable.

In many ways, learning science is similar to becoming fluent in two languages.

One language helps us see how nature behaves. The other helps us describe and predict it with precision.

Together, these languages allow humans to explore and understand the universe in remarkable ways.

19.1 — The Language of Intuition

Long before mathematics entered science, humans understood the world through intuition.

Intuition is the ability to grasp how something works by observing patterns and forming mental pictures of what is happening. It does not always rely on formulas or measurements. Instead, it grows from experience, imagination, and careful observation of nature.

For example, a person can sense that a heavier object is harder to lift than a lighter one without writing any equations. A farmer understands the rhythm of seasons through years of watching the sky, the soil, and the growth of crops. A carpenter knows how wood will behave when shaped or joined with other pieces.

These forms of understanding arise from direct interaction with the world.

In science as well, intuition often appears before formal theory.

Many scientific ideas first emerge as mental models—simple ways of visualizing how nature might behave. When someone imagines planets moving through space, waves traveling through water, or heat spreading through a material, they are using intuitive reasoning.

This kind of thinking allows people to form an initial picture of a phenomenon.

Intuition helps identify patterns, recognize relationships, and suggest possible explanations. It allows scientists and engineers to visualize systems before they are described mathematically.

For example, when thinking about gravity, one might imagine objects being drawn toward each other by an invisible influence. When thinking about electricity, it can be helpful to picture charges flowing through wires like water moving through pipes.

These mental images are not always exact, but they provide a conceptual framework that makes complex ideas easier to grasp.

Intuition therefore acts as the first step in understanding.

It helps the mind organize observations and imagine how different parts of a system interact. Even in advanced science and engineering, intuitive insight often guides the search for new explanations and solutions.

In this way, intuition becomes the first language through which humans interpret nature.

19.2 — The Language of Mathematics

While intuition helps us visualize how nature behaves, mathematics provides a different and powerful way to describe it.

Mathematics is a language built from symbols, numbers, and relationships. Instead of relying on mental pictures alone, it expresses natural laws through precise equations that relate measurable quantities such as distance, time, energy, and force.

This precision makes mathematics extremely valuable in science.

Consider a simple example. Intuition may tell us that objects fall toward the ground. But mathematics allows us to describe exactly how fast they fall and how their motion changes over time. An equation can predict the position of a falling object at any moment, allowing engineers and scientists to calculate trajectories with remarkable accuracy.

Mathematical descriptions also provide a universal form of communication.

While spoken languages vary across cultures, mathematical relationships remain the same everywhere. An equation describing electrical current or gravitational motion has the same meaning regardless of where it is studied.

Because of this universality, mathematics became the primary language of scientific theories.

Scientists use mathematical models to describe complex systems—whether they involve planetary motion, fluid flow, electromagnetic fields, or quantum behavior. These models allow researchers to test ideas, compare predictions with experiments, and refine their understanding of nature.

Mathematics also enables prediction.

Once a natural law is expressed mathematically, it can be used to calculate outcomes under different conditions. Engineers rely on these calculations when designing structures, machines, and technological systems. Without mathematical analysis, predicting how a bridge will handle stress or how an aircraft will behave in flight would be extremely difficult.

In this way, mathematics acts as a tool for precision and reliability.

Where intuition offers a qualitative picture of how something behaves, mathematics provides quantitative detail. It allows us to measure, calculate, and predict phenomena with clarity and consistency.

For this reason, mathematics became an essential language in science and engineering—one that transforms intuitive ideas into precise descriptions of the physical world.

19.3 — Why Mathematics Became Essential in Science

As scientific understanding advanced, researchers began encountering systems that were far too complex to describe using intuition alone.

Simple observations could explain basic phenomena, but predicting how systems behave under different conditions required much greater precision. This is where mathematics became essential.

Consider the motion of planets.

Early observers could track their movement across the sky and recognize repeating patterns. However, predicting the exact position of a planet months or years into the future required precise calculations. Mathematical relationships made it possible to describe these motions accurately and verify them through observation.

The same need for precision appeared in many other areas of science.

In thermodynamics, mathematics allows scientists to calculate how heat flows through engines and materials. In electromagnetism, equations describe how electric and magnetic fields interact and propagate through space. In fluid dynamics, mathematical models explain how liquids and gases move under different pressures and conditions.

Without mathematics, these systems would remain too complicated to analyze clearly.

Mathematics also provides a framework for testing ideas.

A scientific theory is not only a conceptual explanation; it must also produce predictions that can be compared with experiments. When a theory is expressed mathematically, scientists can calculate specific outcomes and then measure whether those predictions match real observations.

If the predictions agree with experiments, the theory gains credibility. If they do not, scientists must revise their understanding.

This ability to test ideas through calculation and measurement is one of the most powerful features of scientific inquiry.

Mathematics also helps scientists uncover hidden relationships within nature.

Sometimes equations reveal patterns that were not immediately obvious through observation alone. For example, mathematical analysis has helped scientists discover new physical laws, predict the existence of particles or phenomena before they were directly observed, and understand complex interactions within natural systems.

In this way, mathematics became more than just a descriptive tool.

It became a central framework for scientific thinking, allowing researchers to analyze complexity, test theories, and predict the behavior of the natural world with increasing accuracy.

19.4 — When Intuition and Mathematics Meet

Although intuition and mathematics may appear to belong to different worlds, the deepest scientific understanding often emerges when both work together.

Intuition provides the mental picture.

It allows scientists and engineers to imagine how a system behaves. When someone visualizes waves spreading through water, charges moving through a conductor, or forces acting within a structure, they are using intuition to build a conceptual model of the phenomenon.

Mathematics then refines this model.

Equations describe the relationships between quantities such as force, energy, velocity, and charge. These relationships allow the intuitive idea to be expressed precisely and tested against experimental observations.

When intuition and mathematics support each other, understanding becomes much stronger.

For example, imagine trying to understand how waves move through water. Intuition might suggest that disturbances travel outward in patterns that spread across the surface. Mathematics can then describe these patterns through equations that determine wave speed, wavelength, and energy distribution.

The intuitive picture helps people visualize what is happening.

The mathematical description allows scientists to calculate and predict the behavior accurately.

Many important scientific breakthroughs occurred through this interaction.

Scientists often begin with an intuitive insight—a new way of imagining how nature might behave. They then develop mathematical models that capture this idea in precise form. Experiments test whether the predictions of the model match reality.

If the predictions succeed, both intuition and mathematics gain credibility.

Even highly mathematical theories usually depend on strong intuitive foundations. Many physicists describe their ideas using mental images—fields stretching through space, waves interfering with one another, particles interacting through forces—before translating those ideas into equations.

The relationship works in the opposite direction as well.

Sometimes mathematical analysis reveals patterns that suggest new intuitive interpretations. Equations may predict behaviors that were not previously imagined, prompting scientists to rethink how they visualize the system.

In this way, intuition and mathematics function like two complementary lenses through which humans understand the universe.

One helps us see the structure of a phenomenon. The other allows us to describe it with precision.

Together, they form a powerful combination that has guided the development of science, engineering, and modern technology.

19.5 — Teaching and Understanding Through Two Languages

If intuition and mathematics are two languages of understanding, then learning science effectively requires becoming comfortable with both.

Many students encounter science first through equations. They learn formulas describing motion, energy, electricity, or waves. While these formulas are essential, they can sometimes appear abstract when the underlying ideas are not yet clear.

On the other hand, intuitive explanations alone may provide a general understanding but lack the precision required to analyze real systems.

Effective teaching therefore often combines both approaches.

An instructor may first describe a phenomenon through observation and visualization. For example, when explaining waves, a teacher might begin with the image of ripples spreading across the surface of water. Students can easily picture how disturbances travel outward in circular patterns.

Once this intuitive picture is clear, mathematics can describe the details of the motion—how wave speed depends on properties of the medium, how wavelength relates to frequency, and how energy is transported.

In this way, intuition prepares the mind to understand the concept, while mathematics allows the concept to be analyzed precisely.

This combination becomes especially valuable in complex fields such as physics, engineering, and mathematics itself. Engineers often visualize how forces move through structures before performing calculations. Physicists may imagine fields or particles interacting before writing equations that describe their behavior.

When both languages work together, learning becomes deeper and more meaningful.

Students no longer see equations as isolated symbols but as expressions of ideas they already understand conceptually. At the same time, intuition becomes more accurate because mathematical relationships clarify how systems behave under different conditions.

For teachers and learners alike, recognizing the importance of both languages can transform the educational experience.

Instead of choosing between intuitive explanations and mathematical rigor, the goal becomes learning how these two perspectives complement one another.

Together they allow humans not only to understand nature, but also to apply that understanding in science, engineering, and technology.

Closing Transition

Throughout this chapter, we have seen that understanding reality often requires two complementary languages.

One is the language of intuition—the mental pictures and conceptual models that help us visualize how the world behaves. This language grows naturally from observation and experience. It allows us to grasp ideas quickly and connect new knowledge with familiar patterns.

The other is the language of mathematics.

Mathematics expresses relationships with precision. It allows scientists and engineers to measure quantities, test theories, and predict outcomes under different conditions. Through equations and calculations, the behavior of natural systems can be described with clarity and reliability.

Neither language alone is sufficient for a complete understanding.

Intuition without mathematics may remain vague or incomplete. Mathematics without intuition may become difficult to interpret or apply. When the two work together, however, they create a powerful framework for exploring the universe.

Intuition suggests how a system might behave. Mathematics verifies and refines that idea. Together they reveal deeper patterns within nature.

This dual approach has guided many of the greatest achievements in science and engineering. It allows people not only to understand natural laws but also to design technologies that apply those laws in practical ways.

As we reach the conclusion of this journey—from early stories to philosophy, science, engineering, and technology—we begin to see a larger picture.

Human progress is not simply the accumulation of facts or formulas.

It is the gradual development of ways of thinking—ways of observing, questioning, imagining, calculating, and building. These abilities allow humanity to understand the universe and shape the world in which we live.

In the final chapter, we will reflect on the broader meaning of this transformation and what it reveals about the evolving relationship between human understanding, knowledge, and the future of civilization.

Chapter 20 — The Journey Continues

The transition is not over. Humanity is still learning to read the universe.

If we look back across the long arc of human history, the journey we have traced in this book begins in a very simple place.

A human sitting outside a cave, looking at the fire, the stars, the storms, and the changing seasons.

At that moment, there were no laboratories, no universities, no scientific theories, and no technological systems. There was only curiosity—an attempt to understand the mysterious world surrounding human life.

From that simple beginning, humanity gradually developed ways of interpreting nature.

Stories were the first language through which people tried to explain the unknown. Myths gave meaning to thunder, lightning, and the movement of the heavens. These stories were not precise explanations, but they represented the first effort to organize experience and understanding.

With time, questioning became deeper.

Philosophers began asking whether events in nature followed hidden patterns or principles. They tried to reason about the structure of the universe, the causes of change, and the relationship between observation and thought.

Science emerged when curiosity combined with careful observation and experimentation.

By measuring, testing, and comparing ideas with reality, humans discovered that nature follows consistent laws governing motion, energy, electricity, light, and matter. These discoveries transformed our understanding of the universe.

Engineering then carried this knowledge into practice.

The principles discovered through science became the foundation for machines, infrastructure, and technologies that define modern civilization. Electricity powers cities, communication networks connect people across continents, and advanced tools allow us to explore environments far beyond the limits of earlier generations.

Yet even after centuries of discovery, an important truth remains.

Our journey of understanding is far from complete.

Despite all the knowledge accumulated so far, the universe continues to present new questions and deeper mysteries. Every discovery reveals new layers of complexity, inviting further exploration.

In this sense, humanity is still learning how to read the universe.

The transition from stories to technology is not a finished chapter in history. It is an ongoing process—one that continues to unfold as new generations observe, question, and explore the world around them.

20.1 — Every Generation Starts Again

Human knowledge grows across centuries, yet every generation begins its journey almost from the beginning.

Children are not born knowing mathematics, science, philosophy, or engineering. They enter the world with curiosity but must gradually learn how previous generations understood nature. Schools, teachers, books, and institutions help transmit this accumulated knowledge from one generation to the next.

In this way, civilization preserves its intellectual progress.

The discoveries of earlier scientists, the inventions of engineers, and the reflections of philosophers become part of a shared heritage. A student studying gravity today benefits from centuries of observation and analysis. An engineer designing modern technology builds upon principles discovered long ago.

Yet inheritance alone is not enough.

Understanding cannot simply be transferred like an object from one person to another. Each individual must recreate knowledge within their own mind. They must observe, question, experiment, and reflect in order to truly grasp the ideas that previous generations discovered.

This is why education is not merely the memorization of facts.

It is a process through which individuals learn how to think, how to ask meaningful questions, and how to connect ideas with real experiences. When students rediscover these ideas for themselves, knowledge becomes alive rather than distant.

History shows that many scientific breakthroughs occurred when individuals looked at familiar ideas from a new perspective.

Even well-established theories may be refined or expanded as new evidence appears or new ways of thinking emerge. Each generation therefore has the opportunity to contribute something new to humanity's understanding of nature.

This continuous renewal keeps knowledge dynamic.

The accumulated wisdom of the past provides a foundation, but curiosity and inquiry ensure that learning never stops. Every new generation enters the same vast universe and begins exploring it again—guided by what came before, yet still discovering new paths of understanding.

20.2 — The Universe Is Still Full of Questions

Despite the remarkable progress humanity has made, the universe still holds far more questions than answers.

Over the past few centuries, science has revealed many important principles governing motion, energy, electricity, light, and matter.

These discoveries have allowed humans to build powerful technologies and understand many aspects of the natural world with great precision.

Yet each discovery has also revealed new mysteries.

When scientists began studying atoms, they eventually discovered that atoms themselves are made of smaller components. Investigating these particles led to the development of quantum physics, which introduced new and sometimes surprising behaviors of matter and energy.

Similarly, studying the motion of planets led to deeper questions about gravity and the structure of space. Exploring the vastness of the universe through telescopes revealed galaxies, black holes, and cosmic phenomena that continue to challenge our understanding.

Even today, many fundamental questions remain open.

What is the full structure of the universe? How did the universe begin and how will it evolve over time? What is the nature of consciousness and the human mind? How do complex systems—from living organisms to planetary climates—maintain stability while constantly changing?

These questions show that scientific knowledge is not a finished collection of answers.

Instead, it is a continuously evolving exploration.

New instruments, technologies, and methods allow researchers to observe phenomena that were previously beyond human reach. Each new observation provides clues that help refine existing theories or inspire entirely new ones.

This ongoing process reflects an important truth.

The universe is not a simple puzzle waiting to be solved once and for all. It is a vast and dynamic system that continues to reveal deeper layers of structure and complexity.

For curious minds, this is not a limitation but an invitation.

Every unanswered question represents an opportunity for future discovery.

20.3 — The Role of Future Thinkers

Every discovery that shaped human civilization began with individuals who were willing to think beyond what was already known.

Throughout history, curious minds have looked at familiar phenomena and asked new questions. They examined ideas that others accepted without doubt and searched for deeper explanations. In doing so, they gradually expanded humanity's understanding of the universe.

Yet these discoveries did not end the journey.

Each generation inherits a body of knowledge developed by earlier thinkers, but the responsibility to continue exploring remains. The future of science, philosophy, and engineering depends on people who are willing to observe carefully, question established ideas, and imagine new possibilities.

Future thinkers will face challenges that previous generations could not have imagined.

New technologies, expanding populations, environmental changes, and increasingly complex global systems will require fresh ways of understanding and solving problems. Addressing these challenges will demand not only technical knowledge but also creativity, ethical judgment, and cooperation across disciplines.

The next scientist may discover new aspects of the universe that reshape our understanding of matter or space. The next engineer may develop technologies that transform how energy is produced or how societies communicate. The next philosopher may offer insights into human consciousness, ethics, or the meaning of knowledge itself.

In many cases, these advances will come from individuals who simply refuse to stop asking questions.

Often the most significant ideas begin as small observations that others overlook. A curious student noticing an unusual pattern in an experiment, or an engineer examining a system that behaves differently than expected, may uncover insights that eventually lead to major breakthroughs.

For this reason, the progress of human knowledge depends on encouraging curiosity and critical thinking.

When individuals feel free to explore ideas, challenge assumptions, and learn from mistakes, innovation becomes possible. Education systems, research institutions, and collaborative communities all play a role in nurturing the next generation of thinkers.

The universe remains open for exploration.

Future scientists, engineers, philosophers, and curious observers will continue the work of understanding the patterns and principles that shape reality. Through their efforts, the human story of discovery will move forward into new and unexpected directions.

20.4 — Technology and Responsibility

As knowledge grows, human capability expands.

Scientific understanding allows us to uncover the principles governing nature, and engineering transforms that understanding into powerful technologies. Over time, these technologies have reshaped nearly every aspect of human life—from transportation and communication to medicine and energy systems.

But with greater capability comes greater responsibility.

In earlier periods of history, human influence on the planet was relatively limited. Small communities interacted with local environments, and technological systems affected only a narrow region. Today, however, modern technologies operate on a global scale.

Energy production influences climate systems. Industrial processes affect ecosystems and natural resources. Digital communication networks shape how societies exchange information and make decisions.

These developments show that technological power must be guided by thoughtful consideration.

Designing a machine or system is not only a technical task; it also involves evaluating how that system will interact with people, societies, and the environment. Engineers, scientists, policymakers, and citizens must consider the broader consequences of technological choices.

Responsible technological development requires balancing several important factors.

Safety must be carefully ensured so that systems do not create unintended risks. Environmental impact must be minimized so that technological progress does not harm the natural systems on which life depends. Social considerations must also be addressed, ensuring that new technologies contribute to human well-being rather than deepen inequality or conflict.

These responsibilities are not obstacles to innovation.

Instead, they guide innovation toward outcomes that support long-term stability and progress. When technology is developed thoughtfully, it can improve health, increase access to knowledge, strengthen infrastructure, and help societies respond to complex challenges.

The same knowledge that allows humans to build powerful systems also provides the ability to reflect on how those systems should be used.

In this sense, technological progress is closely connected with ethical awareness.

As humanity continues learning how to read and apply the laws of nature, it must also learn how to use that knowledge wisely—ensuring that technological power remains aligned with the broader goals of human well-being and sustainable development.

20.5 — Learning to Read the Universe

Throughout this book we have followed the gradual development of human understanding—from stories and philosophical reflection to science, engineering, and modern technology.

Seen from a broader perspective, this entire journey can be understood in a simple way.

Humanity has been learning how to read the universe.

At the beginning, nature appeared as a collection of mysterious events. Thunder, fire, the movement of stars, the growth of plants, and the behavior of animals seemed unpredictable and often frightening. Early stories and myths were attempts to interpret these events and give them meaning.

With time, careful observation revealed patterns.

People began noticing that the seasons followed regular cycles, that celestial bodies moved in predictable paths, and that physical objects behaved according to consistent rules. These patterns suggested that the universe operates according to underlying principles rather than random occurrences.

Science gradually uncovered many of these principles.

Through observation, experimentation, and mathematical reasoning, scientists discovered laws governing motion, energy, electricity, light, and matter. These laws provided a framework for understanding how different parts of the universe interact.

Engineering then transformed this understanding into practical systems.

By applying natural laws, humans developed machines, transportation networks, communication technologies, and energy systems that allow modern civilization to function. The ability to read nature's patterns became the ability to work with them.

Yet reading the universe is not like reading a finished book.

Nature does not present its knowledge all at once. Instead, it reveals its patterns gradually through careful observation and exploration. Each generation learns a little more about the structure of reality and passes that knowledge forward.

In this sense, humanity is still in the process of becoming fluent in the language of the universe.

Every new discovery adds a word or sentence to that language. Every experiment, observation, and technological innovation helps refine our ability to interpret nature's signals.

The journey of understanding therefore continues.

The universe remains open for exploration, and humans continue learning how to read its patterns with increasing clarity and insight.

Closing Reflection

When we step back and look at the entire journey described in this book, one realization becomes clear.

Human progress has never been a straight line.

It began with simple observations of the natural world. Early humans looked at fire, storms, rivers, and the stars with curiosity and wonder. Without formal knowledge or tools, they tried to interpret these experiences through stories and imagination.

Over time, those stories evolved into deeper questions.

Philosophers began asking whether nature followed hidden patterns and principles. Instead of seeing the world as unpredictable, they searched for order and meaning behind natural events.

Science emerged when curiosity was combined with careful observation and experimentation.

Through measurement and reasoning, humans discovered that the universe operates according to consistent laws governing motion, energy, electricity, light, and matter. These discoveries transformed the way people understood reality.

Engineering then carried this understanding into the physical world.

By applying natural laws, humans built machines, structures, and technologies capable of solving complex problems and improving everyday life. Modern civilization itself rests on this ability to translate knowledge into practical systems.

Yet the deeper lesson of this journey is not simply about the accumulation of knowledge.

It is about the human ability to learn, question, and adapt.

Each generation inherits discoveries made by those who came before, but must also rediscover curiosity and understanding for itself. New questions arise, new tools are developed, and new ideas reshape the boundaries of knowledge.

In this sense, the transition described in this book—from stories to philosophy, from science to engineering, and from knowledge to technology—is not a finished chapter.

It is an ongoing process.

Humanity continues learning how to observe the universe, interpret its patterns, and work with its principles. The same curiosity that once led people to watch the stars still drives the exploration of the cosmos, the study of matter, and the development of new technologies.

The journey continues.

And as long as humans remain curious, thoughtful, and responsible in their search for understanding, the story of learning to read the universe will continue to unfold.

Epilogue — The Next Question

The same curiosity that began in caves now explores atoms, galaxies, and consciousness.

At the beginning of this journey, we imagined a human standing outside a cave, watching the fire and the stars.

The world around them was vast, mysterious, and largely unknown. Thunder shook the sky, rivers flowed endlessly toward distant lands, and the stars moved across the night in patterns that no one yet understood. The human mind responded to these mysteries with the simplest and most powerful tool it possessed: curiosity.

That curiosity led to stories.

Stories were the first attempts to give meaning to the unknown. They allowed early humans to organize their experiences and pass their understanding to others. Over generations, those stories slowly transformed into deeper questions.

Why does the sun move across the sky? Why do objects fall to the ground? What causes fire, wind, and rain?

As questions replaced stories, philosophy emerged. Thinkers began searching for order within nature. They suspected that the universe might follow hidden principles rather than unpredictable moods.

Science eventually gave form to this suspicion.

Through observation, experimentation, and measurement, humans discovered that natural events follow consistent laws. Motion, energy, electricity, light, and matter behave according to patterns that can be studied, understood, and predicted.

Engineering then transformed this understanding into practical capability.

The principles discovered by science became the foundation for machines, infrastructure, communication networks, and technologies that shape modern civilization. Humans learned not only to understand nature, but also to work with it.

Yet the most remarkable aspect of this journey is that the original curiosity has never disappeared.

The same impulse that once asked questions about fire and storms now asks questions about atoms, galaxies, and the human mind. Scientists explore the structure of matter at scales far smaller than anything visible to the eye. Astronomers study galaxies billions of light-years away. Researchers investigate the nature of consciousness and the workings of the brain.

Each of these explorations continues the same tradition.

Humans observe something puzzling. They ask questions. They test ideas. Gradually, understanding grows.

And with every answer, new questions appear.

This is the enduring nature of human curiosity. Knowledge does not end inquiry; it expands it. Every discovery opens new areas of exploration and reveals deeper layers of complexity within the universe.

In this way, the journey of understanding never truly reaches a final destination.

Instead, it moves forward through an endless sequence of questions.

The next generation will ask questions that we have not yet imagined. They will observe phenomena that remain hidden today. They will develop tools and ideas that extend the boundaries of knowledge even further.

The cave fire and the distant stars still belong to the same universe.

The human mind that once wondered about them continues to wonder today.

And somewhere, perhaps in a classroom, a laboratory, or under the open night sky, someone is already asking the next question.

The Next Reader

Long ago, a human being sat outside a cave and watched the world with wonder.

Fire flickered in the darkness. Storms moved across the sky. Stars appeared each night in patterns that no one yet understood.

That human asked a question.

From that moment, a long journey began.

Stories tried to explain the mysteries of nature. Philosophy searched for deeper meaning behind those stories. Science tested ideas through observation and experiment. Engineering transformed knowledge into tools and machines. Technology reshaped the world in ways our ancestors could never have imagined.

Across thousands of years, humanity slowly learned how to interpret the patterns of the universe.

Yet the journey has never truly ended.

Every generation inherits knowledge from those who came before, but each must rediscover curiosity for itself. Every new discovery opens the door to deeper questions. The universe continues to reveal its patterns gradually, inviting us to observe, think, and explore.

Somewhere today, a student is looking at the stars and wondering how the universe works.

Somewhere a young mind is asking a question that no one has asked before.

Somewhere a future scientist, engineer, philosopher, or teacher is taking the first step in the same journey that began long ago.

Perhaps that person is the one holding this book.

Because the story of human understanding is not written by the past alone.

It is written by every curious mind that continues to ask the next question.

Author's Note

This book did not begin as a writing project. It began as a moment in a classroom.

One day, while teaching crystallography, a student asked a question that seemed simple at first but carried a deeper meaning:

"Sir, if the universe runs so simply, why do we describe it in such complicated mathematical language?"

The question stayed with me long after the lecture ended. It reflected a common feeling many students experience when learning science. Nature often appears simple and elegant when observed directly, yet the language used to describe it in textbooks can seem complex and abstract.

Over the years, while teaching physics and interacting with students, I realized that this gap between **intuition and formal description** often makes science appear more difficult than it really is.

In reality, human understanding has evolved gradually. Our ancestors first interpreted the world through stories and imagination. Later, philosophers began asking deeper questions about nature. Science emerged when observation and experimentation became systematic. Mathematics then provided a precise language for describing natural laws, and engineering transformed that knowledge into the technologies that shape our civilization.

Seen in this way, science is not a collection of isolated subjects. It is a **long human journey of understanding**.

This book attempts to present that journey in a simple and connected way—from early curiosity to modern technology. Rather than focusing on technical details, it explores how human thinking evolved and how different ways of understanding nature gradually emerged.

As a teacher, my hope is that this perspective helps readers—especially students—see science not as a difficult set of formulas, but as a natural extension of human curiosity.

The same curiosity that once led people to watch the stars and wonder about storms still drives scientific exploration today.

And perhaps somewhere, in another classroom, a new question will begin the next chapter of that journey.

Key Ideas from the Book

The story presented in this book can be understood through a few simple but powerful ideas. These ideas describe how human understanding of the universe gradually evolved over time.

1. Curiosity is the beginning of knowledge. Human civilization began when people started asking questions about the world around them—about the stars, the seasons, fire, life, and death. Curiosity is the force that drives all exploration and discovery.

2. Stories were the first language of understanding. Before science existed, early humans explained natural events through stories and myths. These stories were the earliest attempts to interpret the mysteries of nature.

3. Questions transformed stories into philosophy. As curiosity deepened, humans began asking why things happen rather than only describing them. This transition marked the birth of philosophical thinking about nature.

4. Observation and testing gave birth to science. Science emerged when ideas about nature began to be tested through observation, measurement, and experimentation. This allowed humans to discover consistent laws governing natural phenomena.

5. Mathematics became the precise language of nature. While intuition helps us visualize how systems behave, mathematics allows scientists to describe those behaviors with accuracy and predict outcomes under different conditions.

6. Engineering translates knowledge into action. Engineering applies scientific understanding to create machines, structures, and systems that solve practical problems and improve everyday life.

7. Technology reshapes civilization. The tools and systems built through engineering—from engines and electricity to communication networks and computers—have transformed how societies function.

8. The journey of understanding continues. Despite centuries of discovery, humanity is still learning how to interpret the patterns of the universe. Each generation inherits knowledge from the past and contributes new insights for the future.

Together, these ideas show that human progress is not simply the accumulation of information. It is a continuous process of curiosity, questioning, discovery, and application that allows humanity to gradually learn how to read the universe.

Glossary

This glossary introduces some of the key terms used throughout the book and helps readers connect the ideas presented in different chapters.

Curiosity: The natural desire to ask questions and explore the unknown. Curiosity is the starting point of human learning and discovery.

Story or Myth: An early way of explaining natural events through imagination and narrative before systematic observation and scientific reasoning developed.

Philosophy: A method of thinking that seeks to understand fundamental questions about nature, existence, knowledge, and reality through reasoning and reflection.

Observation: Carefully noticing and recording natural phenomena. Observation is one of the first steps in scientific investigation.

Experiment: A controlled test designed to examine whether an idea or explanation about nature is correct. Experiments allow scientists to verify or challenge theories.

Scientific Method: A systematic approach to understanding nature that involves observation, forming hypotheses, testing through experiments, and refining ideas based on evidence.

Mathematics: A precise language using numbers, symbols, and relationships to describe patterns and laws in nature. Mathematics allows scientists to measure, calculate, and predict phenomena.

Scientific Law: A consistent relationship observed in nature that describes how certain phenomena behave under specific conditions.

Theory: A well-supported explanation of natural phenomena based on evidence from observation and experimentation.

Energy: A fundamental concept in science describing the ability of a system to produce change or perform work.

Engineering: The application of scientific principles and knowledge to design and build machines, structures, and systems that solve practical problems.

Technology: Tools, machines, and systems developed by humans using scientific knowledge and engineering to improve everyday life.

Civilization: A complex human society characterized by organized institutions, shared knowledge, technology, and cultural development.

Discovery: The process of uncovering new knowledge about the natural world through observation, experimentation, and reasoning.

Innovation: The creation of new ideas, methods, or technologies that improve existing systems or solve previously unsolved problems.

How to Read This Book

This book is not a textbook, and it does not attempt to teach science through formulas or detailed technical explanations. Instead, it tells a story—the story of how human beings gradually learned to understand the universe.

The chapters follow the long journey of human thinking. The book begins with early curiosity about nature, moves through philosophy and the birth of scientific reasoning, explores the discoveries that revealed the laws of nature, and finally reaches the engineering and technologies that shape modern civilization.

Each part represents a stage in the development of human understanding.

Readers do not need any special background in science to follow this journey. The ideas are presented in simple language so that students, teachers, and general readers can reflect on how knowledge evolves over time.

Rather than focusing on memorizing information, readers are encouraged to pay attention to the **pattern of thinking** that appears throughout the book. Human progress has always followed a similar path: observation leads to questions, questions lead to ideas, ideas are tested through experience and experiment, and successful ideas eventually become tools that transform society.

For students, the book offers a broader perspective on the subjects they encounter in classrooms. Science, mathematics, and engineering are often taught as separate disciplines, but in reality they are parts of a continuous process of discovery.

For teachers, the book may serve as a reminder that every scientific concept ultimately begins with curiosity. Behind every equation or theory lies a simple question about how the world works.

For general readers, the book invites reflection on humanity's shared journey of learning. The technologies surrounding us today are the result of centuries of questioning, exploration, and collaboration.

As you read, you may wish to pause occasionally and reflect on the ideas presented in each chapter. The goal is not simply to finish the book, but to see the larger picture of how human understanding continues to evolve.

The journey described here is still unfolding, and every curious mind becomes a participant in that ongoing story.

Acknowledgements

Every book carries within it the voices, questions, and experiences of many people. Although a single author's name appears on the cover, the ideas inside often grow through countless interactions with others. This book is no exception.

First, I would like to thank my students. Many of the ideas presented here did not originate in solitude but in the lively environment of the classroom. Curious questions, thoughtful doubts, and unexpected insights from students have often pushed me to rethink familiar ideas and explain them more clearly. In fact, the very inspiration for this book came from a simple question asked during a lecture on crystallography: *"Sir, if the universe runs so simply, why do we describe it in such complicated mathematical language?"* That moment reminded me how powerful curiosity can be.

I am also grateful to the long tradition of scientists, philosophers, and engineers whose work forms the foundation of our understanding of the world. Their observations, experiments, and reflections have gradually revealed the patterns that shape nature and human civilization. Every teacher and writer stands on the shoulders of those who explored before us.

My appreciation also goes to fellow educators who continue to nurture curiosity in classrooms across the world. Teaching is not only about transferring information but about awakening the desire to explore and understand. The spirit of inquiry that teachers cultivate is what keeps the journey of knowledge alive.

Finally, I would like to acknowledge the countless thinkers—known and unknown—who, throughout history, dared to ask questions about the world around them. From early observers of the stars to modern researchers studying the deepest mysteries of matter and consciousness, each generation has contributed a piece to humanity's growing understanding of the universe.

This book is dedicated to that shared human journey of curiosity, learning, and discovery.

About the Author

Sandeep J. Chavan is a physics educator, independent researcher, and author who is deeply passionate about making science understandable and meaningful for students and curious readers. After training as an engineer, he dedicated his career to teaching physics, where he is known for explaining complex scientific ideas through clear intuition and simple reasoning.

Over the years, his classroom interactions with students have shaped much of his thinking about how science should be taught and understood. Rather than presenting science as a collection of formulas, he emphasizes the deeper journey of human curiosity—from observation and questioning to discovery and application.

Through his teaching and writing, he encourages learners to see science not merely as an academic subject but as a natural extension of human curiosity about the universe. His work often connects philosophical reflection with scientific understanding, helping readers appreciate the broader context in which knowledge evolves.

He continues to teach, write, and explore ideas about science, learning, and the evolving relationship between human understanding and the universe.

Disclaimer

This book is intended for educational and reflective purposes. It presents a broad narrative about the evolution of human understanding—from early curiosity and storytelling to philosophy, science, engineering, and modern technology. The goal of the book is to help readers appreciate the intellectual journey through which humanity has gradually learned to interpret and work with the laws of nature.

The discussions in this book simplify many historical and scientific developments in order to make the ideas accessible to students, teachers, and general readers. As such, certain complex topics in the history of science, philosophy, and engineering have been presented in a conceptual and illustrative manner rather than in full technical detail.

This work does not aim to replace formal academic textbooks or specialized scientific literature. Readers interested in deeper technical or historical study are encouraged to consult additional sources and scholarly references.

All opinions and interpretations presented in this book belong to the author and are offered as reflections on the development of human thought and knowledge. They are intended to encourage curiosity, learning, and constructive discussion about science and human progress.

Every effort has been made to ensure that the information presented is accurate and responsibly conveyed. However, the author and publisher assume no liability for any errors, omissions, or interpretations arising from the use of this material.

The purpose of this book is to inspire readers to explore the world with curiosity, critical thinking, and a deeper appreciation for the remarkable journey of human understanding.

Further Reading

From the Chavanian Ecosystem

The ideas discussed in this book are part of a broader exploration of human understanding, science, philosophy, and civilization. Readers who wish to continue this journey may explore some of the following works that examine related themes from different perspectives.

Coherence: When Consciousness Becomes Being

Coherence explores how clarity, intelligence, and balance emerge naturally when awareness aligns with the body and life's rhythms. Moving beyond self-help and belief systems, this book reveals why regulation works where control fails—and how living without interference restores quiet, effortless clarity.

Universal Energy Dynamics (UED)

Universal Energy Dynamics (UED) introduces a consequence-based framework for understanding how systems evolve, accumulate pressure, and resolve imbalance. In this thought-provoking work, Sandeep J. Chavan reinterprets physics, human systems, and global events through a unified lens of constraints, gradients, thresholds, and consequence.

Participation – From Identity to Awareness

Most of us believe we act independently. Yet every decision, identity, and outcome emerges from a network of interactions. Participation by Sandeep J Chavan reveals the hidden structure behind human systems—showing how identity, intelligence, narratives, and awareness arise from participation within a shared field of consequences.

Acceleration - Living in Motion

Why does life feel faster, decisions heavier, and stability harder to find? Acceleration explores how connectivity and constant change reshape modern human experience—and how individuals can remain calm, clear, and balanced while the world continues moving.

Uncertainty – The Pre-Resolved Strain

Uncertainty is usually treated as ignorance, randomness, or error. This book argues otherwise. Across physics, mind, society, and technology, uncertainty emerges as a structural condition of finite systems operating in dynamic equilibrium. A cross-disciplinary reframing that positions uncertainty not as defect, but as architecture.

Chavanism: From Self to Civilization

Chavanism presents a clear structural framework for understanding instability in individuals, institutions, and civilizations. Moving beyond ideology, it explores identity, power, correction, and resilience—offering a disciplined lens for navigating modern complexity through awareness, proportion, and early recalibration.

A Brief History of Intelligence

What is intelligence—really?In an age of artificial intelligence, this book reframes intelligence not as computation, but as a living process grounded in awareness. It explains why machines can extend intelligence but never originate it, and what it means to remain human in an intelligent world.

The War of Thoughts Series

A dialogue-driven exploration of fundamental questions about the universe, intelligence, morality, and human behavior. Through debates with legendary thinkers, the series examines scientific, philosophical, and civilizational ideas in a dynamic and engaging way.

The Human Game

An investigation into human behavior, decision-making, morality, and the complex psychological forces shaping modern society.

The Divine Dilemma

A philosophical exploration of divinity, intelligence, and the observer's role in understanding reality, bridging scientific inquiry and spiritual reflection.

The Original Intent

A study of ancient philosophies and the original human intentions behind spiritual traditions, exploring how early wisdom traditions sought to guide humanity toward harmony and stability.

The Equation of Life: Decoding Dharma for a Harmonious Life

A reflection on human ethics, responsibility, and the principles that help individuals live in balance with society and nature.

Vishwarup Darshan: The Complete Vision of Reality

A broader philosophical work that explores the interconnected nature of reality and the human attempt to comprehend the universe as a whole.

These works collectively explore the relationship between **science, philosophy, human behavior, and the evolving understanding of the universe**. Together they form an expanding body of ideas aimed at helping readers see knowledge as an interconnected journey rather than a collection of isolated disciplines.

Don't miss out!

Visit the website below and you can sign up to receive emails whenever Sandeep Chavan publishes a new book. There's no charge and no obligation.

https://books2read.com/r/B-A-EPGPC-VHTCJ

Did you love *Transition*? Then you should read *Coherence*[1] by Sandeep Chavan!

[2]

What if clarity is not something you achieve—but something that returns when interference stops?

Coherence is not a self-help book, a spiritual manual, or a psychological toolkit. It is a quiet exploration of how life functions when awareness, body, and intelligence are aligned rather than forced.

Drawing from lived observation rather than belief systems, In this book independent researcher and educator Sandeep Chavan, reframes consciousness as an emergent process rooted in biological regulation, not abstract thought. He moves beyond moralized interpretations of sattva, rajas, and tamas, presenting them instead as dynamic states of energy and awareness that continuously reshape human experience.

---

1. https://books2read.com/u/mqdJX9

2. https://books2read.com/u/mqdJX9

Across seven carefully structured parts, the book explores:

How experience emerges before meaningWhy the body is not a container but the first instrument of awarenessHow identity forms through repetition and memoryWhy modern life fragments clarity through chronic overstimulationHow regulation restores what control exhaustsWhat it means to act, work, and lead without inner strain

Rather than offering techniques to improve the self, **Coherence** dissolves the assumption that something is fundamentally wrong. It shows how fatigue, anxiety, emotional volatility, and confusion often arise not from weakness, but from misalignment.

The book culminates in a vision of the *coherent human*—someone who functions with clarity not because they are disciplined, but because they are aligned; not because they strive, but because they listen.

Nothing here needs to be believed.Everything here can be verified through experience.

When coherence is present, nothing needs to be added.When it is absent, nothing can compensate.

Read more at https://www.sandeepjchavan.com/.

www.ingramcontent.com/pod-product-compliance
Lightning Source LLC
LaVergne TN
LVHW100258210926
848480LV00003B/962